EXPAND YOUR HORIZONS

SAVE 50% Special Educational Rate

...arn more about our planet and the worlds beyond in *Earth* ...nd *Astronomy* magazines. Each brings you the latest science ...in a compelling, easy-to-read format.

...cover the hottest news on the planet.

...ur world as you never have before, in *Earth*—the only maga-
...devoted to the science of our planet. *Earth* brings the Earth
...environmental sciences alive with timely news stories, stunning
...ographs and in-depth feature articles. You'll probe an active
...no, discover how the dinosaurs *really* lived, get the facts on
...al warming and much more. You'll even find a field trip article
...ery issue so you can explore
...h's most interesting
...es firsthand.
...y now and
...50% off the
...stand rate:
...ar/6 issues,
...75.

...lore the
...rlds beyond
...th.

...k holes, the Big
...g, life on Mars,
...oding stars—you'll
...ore the most intrigu-
...ideas in the universe
...STRONOMY magazine.

...y month ASTRONOMY covers both the science and hobby of
...onomy. Clearly written features and magnificent photography
...you up to date on the latest in astronomical research and
...e exploration. And easy-to-follow observing guides and star
...ts show you how to explore the night sky with binoculars, a
...cope or just the naked eye. Reply now and save 50% off the
...stand rate: 1 year/12 issues, $17.70.

Wm. C. Brown Publishers has made a special arrangement with Kalmbach Publishing Co. to offer you Earth and Astronomy at 50% off the newsstand price.

Enjoy *Earth* or *Astronomy* at special savings.

Just fill out and mail the attached card and you'll save 50% on a one-year subscription. Or call toll free, 800-533-6644, 24 hours a day.

...ach Publishing Co., Dept. A3061, 21027 Crossroads Circle, P.O. Box 1612, Waukesha, WI 53187-1612
Dept. A3064

SPECIAL EDUCATIONAL RATE

Subscribe and Save 50%

☑ **Yes!** Send me the latest issue of *Earth* and enroll me as a new subscriber. I'll save 50% off the newsstand price with this special educational rate of $11.75 for one year (six issues). I understand that if I am not satisfied for any reason, I may cancel my subscription and receive a full refund for any unmailed issues.

❑ Payment enclosed ❑ Bill me

Special offer – New subscribers only

Name ....
Address..
City..
State .... Zip...........

Payable in U.S. funds.

Earth

E3064E

SPECIAL EDUCATIONAL RATE

Subscribe and Save 50%

☑ **Yes!** Send me the latest issue of ASTRONOMY and enroll me as a new subscriber. I'll save 50% off the newsstand price with this special educational rate of $17.70 for one year (12 issues). I understand that if I am not satisfied for any reason, I may cancel my subscription and receive a full refund for any unmailed issues.

❑ Payment enclosed ❑ Bill me

Special offer – New subscribers only

Name ....
Address..
City..
State .... Zip...........

Payable in U.S. funds.

ASTRONOMY

E3064A

Discover the Earth...and sky

50% OFF

Reply now to take advantage of this special educational discount.

Earth

The only magazine devoted to the science of our planet.

ASTRONOMY

The world's most popular astronomy magazine.

 Kalmbach Publishing Co. guarantees that you may cancel your subscription at any time and receive a full refund for any unmailed issues.

Or call 800-533-6644, 24 hours a day

BUSINESS REPLY MAIL
FIRST CLASS MAIL   PERMIT NO. 16   WAUKESHA, WI

POSTAGE WILL BE PAID BY ADDRESSEE

Earth
P. O. Box 1612
Waukesha, WI 53187-9950

NO POSTAGE NECESSARY IF MAILED IN THE UNITED STATES

BUSINESS REPLY MAIL
FIRST CLASS MAIL   PERMIT NO. 16   WAUKESHA, WI

POSTAGE WILL BE PAID BY ADDRESSEE

ASTRONOMY
P. O. Box 1612
Waukesha, WI 53187-9950

NO POSTAGE NECESSARY IF MAILED IN THE UNITED STATES

THE SCIENCE OF OUR PLANET

A Wm. C. Brown Publishers'
Earth Science Reader

Kalmbach Publishing Company

David Dathe
Alverno College

WCB **Wm. C. Brown Publishers**
Dubuque, Iowa • Melbourne, Australia • Oxford, England

Book Team

Executive Editor *Jeffrey L. Hahn*
Developmental Editor *Mary Hill*
Production Editor *Audrey Reiter*
Designer *Jeff Storm*
Photo Editor *Carrie Burger*
Permissions Coordinator *Karen L. Storlie*

Wm. C. Brown Publishers
A Division of Wm. C. Brown Communications, Inc.

Vice President and General Manager *Beverly Kolz*
Vice President, Publisher *Earl McPeek*
Vice President, Director of Sales and Marketing *Virginia S. Moffat*
National Sales Manager *Douglas J. DiNardo*
Marketing Manager *Christopher T. Johnson*
Advertising Manager *Janelle Keeffer*
Director of Production *Colleen A. Yonda*
Publishing Services Manager *Karen J. Slaght*
Permissions/Records Manager *Connie Allendorf*

Wm. C. Brown Communications, Inc.

President and Chief Executive Officer *G. Franklin Lewis*
Corporate Senior Vice President, President of WCB Manufacturing *Roger Meyer*
Corporate Senior Vice President and Chief Financial Officer *Robert Chesterman*

Cover: Lake Louise occupies a classic glacial valley formed by Victorian Glacier, Banff National Park, Alberta, Canada © Doug Sherman/Geofile

Copyright © 1994 by Wm. C. Brown Communications, Inc. All rights reserved

A Times Mirror Company

Library of Congress Catalog Card Number: 93–73080

ISBN 0–697–23408–8

No part of this publication may be reproduced, stored in a retrieval system, or transmitted, in any form or by any means, electronic, mechanical, photocopying, recording, or otherwise, without the prior written permission of the publisher.

Printed in the United States of America by Wm. C. Brown Communications, Inc., 2460 Kerper Boulevard, Dubuque, IA 52001

10 9 8 7 6 5 4 3 2 1

Contents

From the Publisher   1

Part One: Environmental Geology

Monitoring the Global Environment by *Steven A. Zaburunov*   2
    We can use satellites as trouble shooters to scrutinize our ever-changing environment and to assess the health of our planet.

History Lessons by *Elizabeth Culotta*   10
    To learn how global warming may affect Earth's climate, climatologists are looking for clues in the record of geologic history. But can we trust the dusty crystal balls of the past to accurately predict our planet's future?

Under the Ozone Hole by *Shawna Vogel*   14
    Scientists studying the effects of increased ultraviolet radiation on Antarctic life paint a less gloomy picture than expected but still warn of possible damages to marine life.

Oil: When Will We Run Out? by *David G. Howell, Kenneth J. Bird, and Donald L. Gautier*   20
    The theory of plate tectonics is helping geologists to strengthen their predictions of how much oil remains to be discovered. The authors argue that at the current rate of consumption, supplies would last 70 years.

Powder River Coal by *Doug McInnis*   28
    Low-polluting coal in Wyoming and Montana has geologists fascinated and stumped. No one knows how it formed. But the coal has also aroused protests from environmentalists concerned about strip mining.

Part Two: Meteorology

Live from Space: Clouds in the News by *Steven A. Zaburunov*   34
    From their all-encompassing vantage point, shuttle astronauts have a better view to understand Earth's weather systems.

Receive Satellite Images on Your Computer by *Phillip J. Imbrogno*   42
    For less than $1,000, you can receive satellite photos of Earth on your own computer system.

Part Three: Oceanography

Swept Away by *John Dvorak and Tom Peek* — 50
Hundreds of millions of people live in areas vulnerable to tsunamis. Are they ready for the next big wave?

Charting Earth's Final Frontier by *Tom Yulsman* — 58
The *Magellan* spacecraft has mapped 99 percent of the surface of Venus, yet more than 90 percent of the deep ocean bottom remains uncharted. Now new technologies are offering startling images of the undersea landscape.

Part Four: Plate Tectonics

Anatomy of a Mountain Range by *Berkley Chew* — 64
Learn how the Rockies formed on a top-of-the-world trip through rugged southwest Colorado.

Lessons from Landers by *Richard Monastersky* — 72
Seismologists thought they understood big earthquakes fairly well. Then the fourth-largest quake in California history struck the small desert town of Landers and shook their assumptions.

CAT Scanning the Earth by *Jim Dawson* — 80
Supercomputers are enabling geophysicists to visualize the Earth's churning interior and gain new insights into how it works.

Hawaii's Volcanos: Windows into the Earth by *John Dvorak* — 86
In Hawaii, geologists can look the fires of hell in the face and come away with priceless information about the deep Earth.

Part Five: Paleontology

Bakker's Field Guide to Jurassic Park Dinosaurs by *Bob Bakker* — 95
Up-to-date, authoritative information available on the browsers, bashers, and slashers you'll see in the movie.

Extinctions—or, Which Way Did They Go? by *Steven M. Stanley* — 106
Wiping out much of the life on Earth wasn't quite as easy as some astronomers propose.

Part Six: Cartography

Building a Better Map by *Richard J. Pike and Gail P. Thelin* — 116
Digitally-shaded relief maps reveal subtle features not visible on conventional maps.

From the Publisher

Wm. C. Brown Publishers is pleased to offer our customers this exceptional earth science reader. *Earth: The Science of Our Planet—A Wm. C. Brown Publishers' Earth Science Reader* is our first joint venture with Kalmbach Publishing Company, publisher of *EARTH,* a bimonthly magazine.

EARTH is one of the finest earth science magazines available today. It offers its readers spectacular photography and a wide variety of articles written by some of today's best earth science writers. Wm. C. Brown Publishers is proud to offer you the opportunity to enjoy the best of *EARTH* magazine. We have carefully selected sixteen of the most interesting and contemporary articles for you to enjoy. We encourage you to order this reader for your students and to use it in your introductory geology classroom. To help you do this, we have developed an *Instructor's Resource Manual.* This manual, written by David Dathe of Alverno College, is loaded with teaching tips, quizzing questions, and detailed summaries for each article in the reader.

Wm. C. Brown Publishers would like to thank Robert A. Maas, Rhoda Sherwood, and Tom Yulsman from *Earth: The Science of Our Planet.* It was a pleasure working with them on this project.

We trust you will enjoy reading this exciting and timely reader. Please feel free to drop us a line and let us know what you think of *Earth: The Science of Our Planet—A Wm. C. Brown Publishers' Earth Science Reader.*

Jeffrey L. Hahn
Executive Editor

MONITORING THE GLOBAL ENVIRONMENT

by Steven A. Zaburunov

One of the most comprehensive ways to study our own fragile environment is from space.

The application of space-based technology for global monitoring is threefold: mapping landforms, detecting trends in changing climates and vegetation, and analyzing spectra of the physical and/or chemical properties of vegetation, rocks, and the atmosphere.

Although getting a space perspective is expensive, the pristine view from orbit offers significant benefits that easily outweigh the costs. A study that would take an ocean-going research vessel years to complete, a polar orbiting satellite can do in about 72 hours. The same satellite can glide over stormy or icy seas that the ship could never safely navigate. One satellite can also do the job of an army of weather balloons or of many high-flying airplanes.

Satellite data can be more objective and detailed than ground-based measurements. If there is a region that the satellite does not cover, that region will appear blank on the finished map. By contrast, ground-based measurements may sometimes be interpolated across most areas of missing data.

There are many ways our environment can be monitored from a satellite platform; the following pages show just a few. When it comes to monitoring our home world, including our oceans, atmosphere, and land vegetation, satellites are invaluable tools.

The satellites beam their data back to receiving stations located all over the planet. The data are sent to a central processing office, where they are computer-processed to be interpreted. Data beamed from satellites improve our understanding of the ozone layer and help keep us aware of damage done to this layer. Computer-enhanced images of our thinning blanket of ozone are instrumental in convincing many countries to modify manufacturing processes.

Remote sensing, as it is generally called, can monitor an amazing variety of our planet's major health indicators. By measuring the spectra of light returning from a forest, researchers can determine its health, the population density of the trees, and the variety of trees. When forest fires broke out in Yellowstone National Park in the summer of 1988, satellite images helped estimate the true extent and impact of the blazing fire from the coolness of space.

Researchers at Goddard Space Flight Center plan to establish a satellite system to monitor our global environment. The project may be operational as soon as 1998. The most exciting prospect is that the system could discover global relationships that no one has been looking for.

Our depleting ozone layer is cause for global concern. The extra ultraviolet light allowed in by the disappearing layer could seriously harm animals and vegetation, including agricultural products. The Nimbus-7 image above, taken of the South Pole October 5, 1991, shows ozone levels at the lowest level measured to date. (NASA/GSFC)

*T*he fires in Yellowstone National Park in 1988 are in their early stages in this false color image. Burned vegetation shows red and purple; smoke from active fires is light blue and white; and healthy vegetation is green. Darker green represents dense stands of timber. (EROS Data Center)

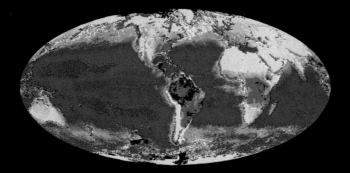

The Global Biosphere
(green = lush vegetation, yellow = desert)
NASA/GSFC

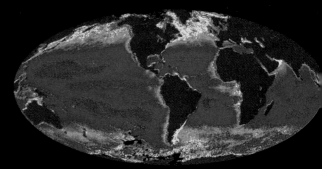

Phytoplankton Pigment Concentration
(green = good fishing, red = bad fishing)
Coastal Zone Color Scanner
Nimbus-7

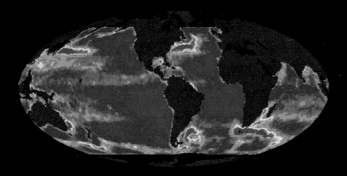

Sea Level Variability
(red = highly variable current direction, blue = steady current direction)
Geosat Altimeter
Geosat

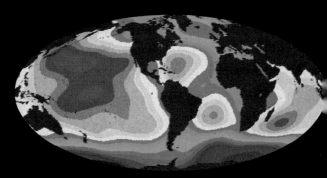

Dynamic Topography
(Large-scale average ocean circulation rate;
red = center of circulation pattern)
Geosat Altimeter
Geosat

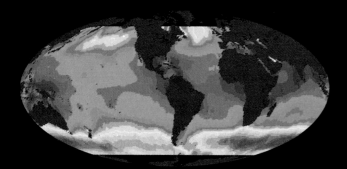

Significant Wave Height
(red = roughest sea, blue = most peaceful seas)
Geosat Altimeter
Geosat

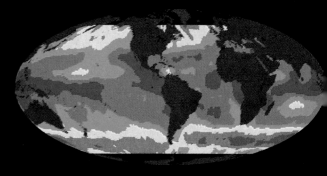

Mean Wind Speed
(red = most intense winds, blue = calmest winds)
Geosat

Normalized Difference Vegetation Index
(green = lush vegetation, yellow = desert)
Advanced Very High Resolution Radiometer
NOAA-7

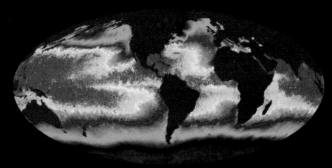

Sea Surface Temperature
(blue = cold, red = hot)
Advanced Very High Resolution Radiometer
NOAA-9

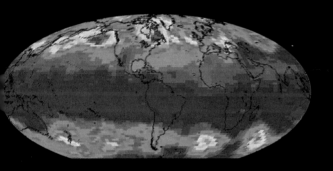

Total Ozone Concentration
(For all layers of the atmosphere; purple = highest, red = lowest)
Total Ozone Mapping Spectrometer
Nimbus-7

Snow Depth and Sea Ice Coverage
(purple = deepest snow, white = sea ice)
Scanning Multi-channel Microwave Radiometer
Nimbus-7

The variety of information that satellites detect increases constantly. The proposed Earth Orbiting System (EOS) could monitor water and energy cycles, the chemistry of the upper atmosphere, sources and sinks of greenhouse gases, and changes in land cover. The growth or decline of glaciers and polar ice sheets will allow predictions of changing sea levels and global water balance. EOS can watch volcanos and observe their role in global climate change.

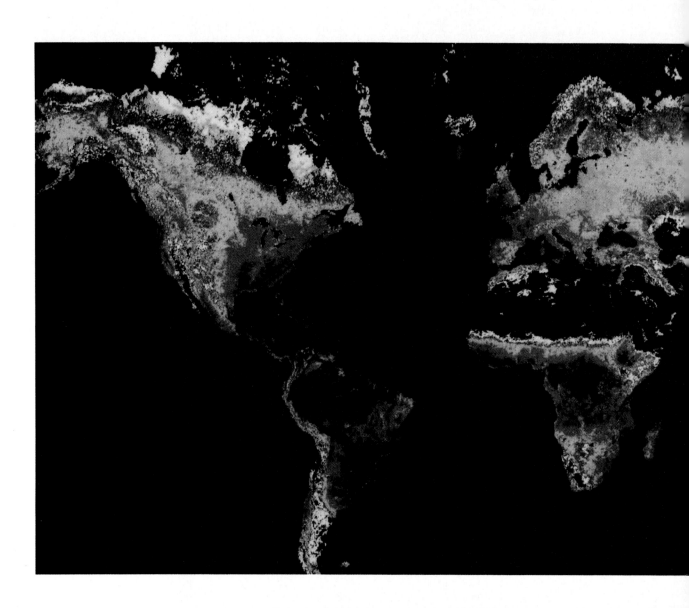

*T*his global image (left) indicates vegetation levels in all parts of the world. Named the Global Vegetation Index, it provides a false color image indicating land areas with the potential for plant life. The colors are arranged from brown for barren land to bright green for lush tropical forests. (NOAA/NEDIS, 1987, EROS Data Center)

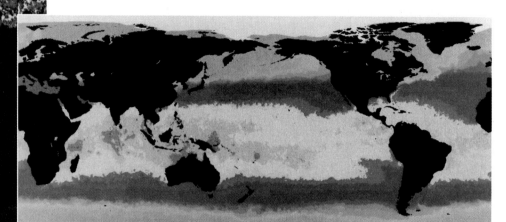

*T*hree global images (right) were taken with microwave sensors from polar orbiting satellites. The satellites easily map surface temperatures (top) from an orbiting platform. They can also map the amount of water vapor in the air (middle) as opposed to the amount of water (bottom). (EROS Data Center)

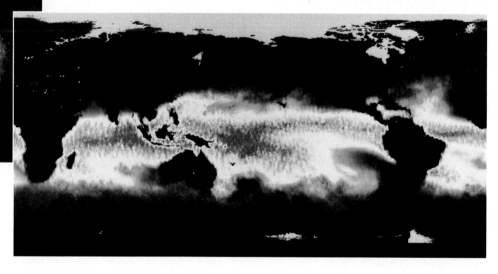

*V*egetation, green to our eye, shows in shades of red in the image at left taken with infrared film. The mosaic is a composite of 15 images taken on clear days from May 24, 1984, through May 14, 1986. (EROS Data Center)

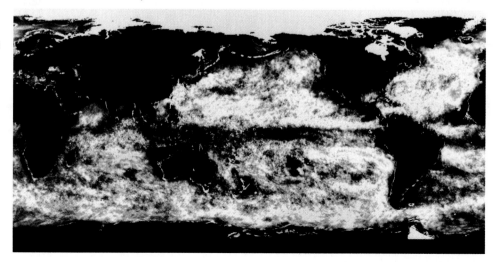

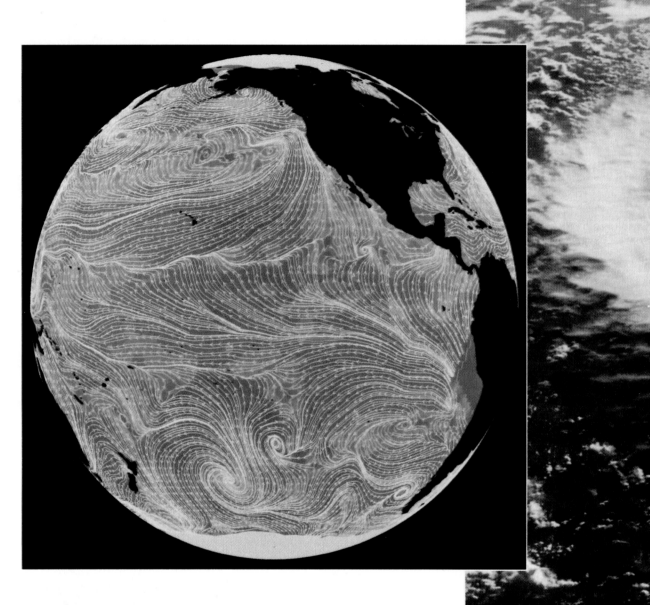

*S*atellites can trace ocean winds blowing along the surface of the sea hundreds of miles away. In the image above, the colors represent wind speed and the arrows indicate direction. Blue indicates wind speeds of 2 to 9 mph, gray from 9 to 13, red from 13 to 35, and yellow, a roaring 35 to 43 mph. Notice the wind speeds near storms in the South Pacific as well as near the coast of Alaska. This swirling image was taken as part of the Seasat Scatterometer Analysis program on September 14, 1978, at 1800 hours GMT. (NASA 89-HC-275)

*S*pace shuttle photography allows us to see weather phenomena in their entirety. This photo of Typhoon Pat showing its full fury was taken from Discovery on August 20, 1985. (NASA 85-HC-345)

Hurricane Hugo reveals its secrets to a satellite traveling high above its powerful winds. Taken September 22, 1989, the image at left illustrates the convenience of satellite monitoring. The only other way to gather such data is to fly through the dangerous winds in a sturdy plane. (EROS Data Center)

EARTH BEAT

History lessons:
Past climates hold clues to global warming

by Elizabeth Culotta

Every student in Earth Science 101 learns some version of geology's first principle: The present is the key to the past. But today, given the possibility of global warming from increasing carbon dioxide (CO_2) in the atmosphere, scientists also follow an inverse corollary to that axiom. They believe Earth's climatological past may hold the secrets of the present and perhaps even the future.

In the geologic past, Earth has been both much colder and much warmer than it is now. Indeed our planet has experienced climate extremes as dramatic as any in science fiction. For example, during the Cretaceous Period, between 65 million and 100 million years ago, breadfruit trees thrived in western Greenland, above the Arctic Circle. Yet only 18,000 years ago, at the peak of the most recent Ice Age, the northeastern United States shivered under ice a mile thick. What caused Earth's thermostat to rise and fall so dramatically? And what can the long-buried past tell us about future climate?

Climatologists use an array of methods to extract information on past climates from the record left in rocks and ice. Perhaps the most straightforward approach comes from scientists in Russia. They argue that past climatic conditions associated with higher levels of carbon dioxide will repeat themselves as the amount of atmospheric carbon dioxide continues to climb. In other words, past climates are direct harbingers of things to come.

Other scientists are wary of extrapolating so directly from past to future. Climatologists in the United States, for instance, generally use computer models of the climate system to predict the future. But they do rely on the geological record to test the accuracy of the models. They enter information on factors known to have influenced past climates, such as carbon dioxide levels, and then see how accurately the models simulate previous epochs. And both the Russians and U.S. scientists have tried a third approach to using past climate, one that pushes the limits of geological evidence: They use data from eons past to measure precisely how much Earth warms when carbon dioxide in the atmosphere increases by a given amount. Many climatologists think this approach is worth trying, but they disagree sharply on whether the data are accurate enough to predict future temperatures reliably.

The record of past climates is by no means easy reading. But the geological evidence supporting the central tenet of global warming theory is clear: In the past, when there was more carbon dioxide in the atmosphere, Earth was warmer. This warming was due to the greenhouse effect. Much like the clear panes of glass in a greenhouse, carbon dioxide and certain other gases in the atmos-

rings turn the desert green here in the
wa Oasis in Egypt. Increased rainfall
the now parched Sahara resulting from
e greenhouse effect could make scenes
ke this more common.

here allow the sun's energy to pass
hrough in the form of light. But
hen some of that energy reradiates
om the Earth as heat, these greenhouse gases trap the heat in the atmosphere, thus warming the planet.

Scientists agree that the greenhouse effect is real; without it, Earth's
average temperature would be about
0 degrees Fahrenheit colder than it is
ow. And over the past 150,000 years,
ases trapped in ice cores show a remarkable correspondence between
igh CO_2 levels and high temperature, although other factors, in addition to carbon dioxide, have also
aused temperature change.

How much Earth will warm due
o increasing CO_2 is only part of the
quation, though. Climate encompasses not only temperature, but also
ainfall, wind patterns and a host of
ther factors. Russian scientists have
xplored the connection between carbon dioxide and various climatic conditions through what they call the
paleoanalog method," developed
rimarily by climatologist Mikhail
Budyko of the State Hydrological Institute in St. Petersburg in the mid-
970s. At that time, few in the West
ad studied the past to predict future
limates. Now even Western scientists skeptical of Budyko's methods
dmit that he was a pioneer.

Budyko's method was simple and
made intuitive sense. First, he and his
olleagues gathered an exhaustive array of data from both the distant and
he more recent past. They sought
lues to air temperatures, the amount
f carbon dioxide in the atmosphere,
he amount of water in the soil and so
on. They inferred temperature and
ainfall millions of years ago, for example, from spores and pollen, which
evealed the kinds of plants that had
ived and died in various regions.
They gathered additional information
n rainfall from sedimentary deposits
on lake shores, which revealed how
ake levels had fluctuated over
housands of years. And like their

counterparts elsewhere, the Russians
inferred temperatures from the faint
chemical signatures left in fossils and
muds on the ocean floor.

The researchers compiled all the
data and created maps of ancient
temperature and rainfall. Because
their main concern was predicting the
possible climatic effects of global
warming, they concentrated on periods when the climate was warmer
than it is now. For example, during
the Early Pliocene, about 3.8 million
years ago, the Russians estimated that

Our planet has experienced climate extremes as dramatic as any in science fiction.

---

there was more than twice as much
CO_2 in the atmosphere as there is today, and that temperatures were
about 5 F higher. They also found evidence of heavy precipitation in some
northern continents. According to the
paleoanalog method, climatic conditions similar to these will reappear
when CO_2 levels double sometime in
the middle of the next century, says
Andrei Lapeneis, a Russian-trained
paleoclimatologist now working at
New York University.

Many in the United States disagree
with the Russians' conclusions. They
don't quarrel with the notion of reconstructing past climates, of course.
In fact, both groups of scientists often
agree on the main features of Earth's
climatic history. But they split over
the idea that the future will imitate
the past, says Tom Crowley, a paleoclimatologist at the Applied Research
Corporation in College Station,
Texas. He and other researchers say
that even if the global average temperature does climb 5 F, there's no
guarantee that details of climate, such
as rainfall and regional temperatures,
will be the same in the future as they
were in the past.

Also, as many climatologists point
out, some past climatic changes were
probably driven by forces other than
carbon dioxide. Thus, skeptics of the
paleoanalog method argue that predictions of future climates based
solely on carbon dioxide levels may
be misleading.

For example, most scientists believe that the Northern Hemisphere
experienced hotter summers and
colder winters 9,000 years ago — not
because of the greenhouse effect but
because of slight differences in
Earth's orbit and the tilt of its axis.
The combination of these factors
caused greater seasonal extremes of
hot and cold. Today, with a slightly
different orbital configuration and
axial tilt, the seasonal changes are less
extreme.

Also, Earth's climate has often
changed slowly in the past, over
thousands of years, while the industrial release of greenhouse gases has
occurred in the blink of an eye, geologically speaking. The climate's response to today's rapid change may
differ from its response to the slow
changes in the past, Crowley and others argue.

"There's no obviously identical
period in the past," says paleoclimatologist John E. Kutzbach of University of Wisconsin-Madison. "We
don't repeat ourselves in time."

For this reason, Kutzbach has
chosen to work with computer
models to predict future climate. He and his colleagues input
likely future conditions, such as an
increase in atmospheric carbon dioxide and ask the computer to simulate
the resulting climate. But to test the
models, they re-run the programs using the geography and orbital conditions known to have prevailed in
previous epochs. If the computer
model reconstructs the past climates
accurately, the climatologists feel relatively confident in its ability to predict future climate change.

Beginning in the late 1970s,
Kutzbach helped direct a research effort called the Cooperative Holocene
Mapping Project (COHMAP) to explore how changes in Earth's orbit
have affected climate from 18,000
years ago until today. In many ways,

EARTH BEAT

the world 18,000 years ago was similar to today's: The continents, oceans and mountain ranges had all arrived at roughly modern positions. But there were differences in the tilt of the Earth's axis and the configuration of its orbit.

An international team of COHMAP members fed information on these differences, along with some basic geological data, into their computer model. They began their simulation at 18,000 years ago and ran it until the present. Then they checked the model's results against the climatic picture sketched by geologic evidence. In many cases, data and model agreed. For example, 9,000 years ago — at the height of the orbital change — the model found hotter summers and colder winters in the northern hemisphere and stronger monsoons in northern Asia. The model even produced a lush, wet climate complete with lakes in what is now the Sahara Desert, just as the geological data show.

Yet despite these promising results, computer models are far from perfect. Crowley created what he calls a "paleo-modeling scorecard," and while the COHMAP model received a "very good," the scorecard described many other trials as "fair" or "poor." Also, atmospheric modelers have trouble recreating the most striking feature of the past three million years: the periodic advance and retreat of the ice ages. When scientists input changes in Earth's orbit and axis traditionally thought to trigger glacial periods, the models don't show an ice age.

So while climate modelers perhaps have valid criticisms of the paleoanalog method, they can't claim complete success in predicting climate change either. Meanwhile, both groups of scientists have used Earth's history in yet another way. They seek to extract a crucial value, called climate sensitivity, which indicates how much the global average temperature will rise when carbon dioxide increases by a given amount. This number is typically calculated for a doubling of CO_2, and is something of a Holy Grail for climatologists. Scientists already know that atmospheric carbon dioxide has increased more than 30 percent from pre-industrial values. The amount of CO_2 in the atmosphere is expected to double pre-industrial levels sometime in the middle of the next century. The billion-dollar question is, how much warming will the extra CO_2 generate?

Various computer models have estimated climate sensitivities that range from about 4 to about 9 F. That means if the amount of CO_2 in the atmosphere doubles, the global average

> **Past climates contain all the information scientists need to measure how much temperature changes when the amount of carbon dioxide changes.**

temperature should rise somewhere between 4 and 9 F. But the range is very wide: The climate of a world that's five degrees warmer could be drastically different than the climate of a world that's only two degrees warmer. After all, the global average temperature during the last ice age was only about 5 to 10 F colder than the present.

Current climate models give this broad range of results because they don't include all key feedbacks that influence climate, like clouds. If the additional CO_2 warms the climate, for instance, the planet will probably be cloudier. And clouds themselves can affect the temperature in complex ways. Some kinds of clouds can trap heat inside the atmosphere. Other kinds can reflect sunlight back into space, which would have an overall cooling effect. No one is sure whether more clouds would increase the warming or slow it down, and the models reflect this uncertainty.

This is another reason data from the past can be so useful. Clouds and all other such complicated feedbacks were naturally incorporated into the system as Earth warmed and cooled over the eons, says climatologist James Hansen of the NASA Goddard Institute for Space Studies in New York City. Thus past climates contain all the information scientists need to measure how much temperature changes when the amount of carbon dioxide changes. In 1990, Hansen himself calculated climate sensitivity by using the best estimates of Earth's temperature and atmospheric carbon dioxide in the last ice age, 18,000 years ago. He concluded that if carbon dioxide levels double, the global average temperature will rise about 5.4 F, plus or minus about two degrees. Russian paleoanalog reconstructions of warmer periods also give a figure of about 5.4 F, although they do not provide a range of uncertainty. Most recently, Martin Hoffert of New York University and colleague Curt Covey of the Lawrence Livermore National Laboratory used both the chilly glacial maximum and the steamy Cretaceous to calculate a value of about 4 to 5.4 F.

Some scientists have criticized these last results, because estimates of the amount of CO_2 in the Cretaceous atmosphere are fuzzy at best. Then again, other scientists are comforted that the results of these studies fall within the range predicted by the computer models.

If nothing else, all these examples confirm what geology students are likely to learn by the time they reach Earth Science 404: That old adage about past and present is not always true. Many events in Earth history have no modern analog, just as our release of greenhouse gases has no parallel in the past. Thus, eager scientists may risk over-interpreting an ambiguous geological record. But climatologists seeking a vision of the future are likely to expend more and more effort peering through the dusty crystal ball of the past. For actual data on what happens when the climate changes, there's simply no place else to look. ⊕

Under the

*From orbit, the atmosphere looks like a glowing blue shield betwe
layer has been breached. Scientists are only now beginning*

Photo courtesy NASA

ozone hole

...rth and space. Within that shield, however, the invisible ozone ...derstand the consequences for the creatures that live below.

By Shawna Vogel

After several months of perpetual darkness, spring rolls in over the icy Antarctic like the dawning of a new day. The sun sits above the horizon a few moments longer each afternoon and polar plants and animals begin shaking off the deep winter cold.

But during these dim spring days, a new and dangerous kind of light shines down on the South Pole, for this is when the ozone hole opens over the Antarctic. All winter long a vortex of wind has whipped around the pole, isolating the air mass above it, and the bitter cold has fostered a layer of icy clouds high up in the polar stratosphere. On the surfaces of the ice particles that make up these clouds chlorine nitrate reacts with other chemicals to produce free molecular chlorine. Once the springtime sunlight hits, that chlorine — much of it from chlorofluorocarbons and hydrochlorofluorocarbons in refrigerants, aerosol sprays and solvents — begins to devour ozone, carving a hole in the ozone shield that screens out damaging ultraviolet radiation. An invisible shaft of ultraviolet radiation then streams through the gap.

According to a recent report from the United Nations Environment Programs, the ozone hole has steadily widened since it first opened above Antarctica during the late 1970s. Researchers are now seeing signs of major ozone depletion over the North Pole as well. In 1992, only a last-minute rise in temperatures prevented Arctic ozone losses comparable to those in the Antarctic ozone hole. Different air circulation patterns in the Arctic will probably prevent ozone losses there from taking the form of a well-defined "hole." But severe ozone depletion may hit the north soon — if not this spring, then within the decade, says Joe Waters, who heads a project at NASA's Jet Propulsion Laboratory that is using satellites to study atmospheric chemistry. Indeed, the layer of ozone is thinning noticeably at most latitudes, and during the last decade the annual dose of harmful UV striking the northern hemisphere rose by about 5 percent, according to the U. N. report.

Humans are among the creatures most vulnerable to this downpour of radiation. Unprotected by scales or feathers, we face a rising threat of skin cancer, cataracts and possible immune system deficiencies. But people can avoid UV rays and be treated for the harm they cause. Other animals and plants cannot. In the last few years, researchers have begun to piece together the impact that increasing levels of UV radiation may have on Earth's ecosystems. And recent observations in the field suggest that although a catastrophe is unlikely, global ecosystems will not go unscathed.

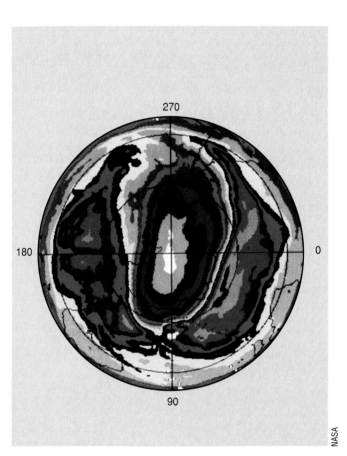

The Total Ozone Mapping Satellite observed the lowest ozone levels ever in the Antarctic ozone hole on October 6, 1991. White and pink represent the areas of greatest depletion.

Researchers' efforts have focused on Antarctica, where the ozone layer reaches its thinnest point. Here, little life exists above the water line. Penguins, seals and seabirds do venture onto land, but they mainly do so only to breed. When they are on land and at risk of high UV exposure, they are effectively shielded by their feathers and fur. Their eggs, in addition to being protected by the nesting parent's body, have shells that are opaque to UV. The eyes and noses of these creatures are vulnerable to the increasing radiation, however, and penguins staring at the sky may develop more eye and skin disorders.

With so few terrestrial species in the Antarctic, the main concern is for marine life. UV rays penetrate the glassy southern ocean to depths of several yards. As a result, the creatures that stand to lose the most are those that live near the surface, predominantly phytoplankton, says environmental biologist Deneb Karentz, of the University of San Francisco.

Phytoplankton are microscopic plants that carpet the surface of the ocean like grass on a plain. They serve as tiny food factories, continuously converting water and the sun's energy into carbohydrates. Phytoplankton are at the bottom of the food chain, which means that they are vital to all sea animals, from tiny krill and zooplankton to large baleen whales that depend on these surface drifters as their ultimate source of energy. The rate of growth of phytoplankton thus sets the productivity of the entire ecosystem.

For that reason, phytoplankton have become minor celebrities in scientific studies of ozone depletion. Since the ozone hole in the Antarctic was discovered, scientists have worried that increased UV might devastate phytoplankton productivity, with some researchers predicting as much as a 40 percent decline in the microscopic plants.

Such a dramatic drop in growth would reverberate throughout the marine ecosystem, perhaps to the point that countries relying on the sea for the bulk of their protein could suffer economically. A substantial decrease in phytoplankton growth could also intensify global warming. Phytoplankton take up carbon dioxide, the greenhouse gas, just the way land plants do. About half of the CO_2 we spew into the atmosphere each year actually is removed from the air by natural processes, and researchers estimate that more than half of the carbon dioxide removed is absorbed by phytoplankton.

But since these fears were first expressed, researchers have scaled down the initial estimates of phytoplankton losses. During the 1990 ozone hole season, geographer Ray Smith of the University of California in Santa Barbara and a group of his colleagues voyaged to the Antarctic to measure the effects of elevated UV on these creatures. They sailed into the Bellingshausen Sea just at the edge of an ice pack where phytoplankton blooms vigorously in austral spring. The ozone hole swept back and forth over the researchers with the shifting winds of the polar vortex. Moving in and out of this intensified UV beam, they were able to compare the effects of the elevated radiation against normal springtime levels.

During the six weeks of the cruise, the team found a 6 to 12 percent drop in the growth rate of phytoplankton exposed to increased UV radiation. But averaged over a full year, the loss in productivity was estimated to be only 2 to 4 percent because the ozone hole was open for only three of the twelve months. Taken against a background variation of 25 percent due to normal shifts in weather and other factors, the results seem far less worrisome than researchers' original estimates.

According to team member Robert Bidigare, a biological oceanographer from the University of Hawaii in Honolulu, the observed drop in phytoplankton might be significant in a harsh Antarctic system already limited in food. "But what's not happening is you're not having the phytoplankton being decimated as earlier reports claimed."

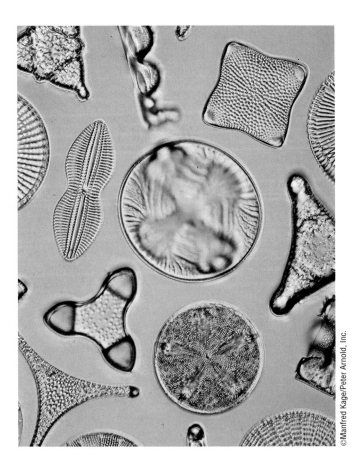

Diatoms' silica-containing shells give the single-celled organisms a jewel-like appearance. Diatoms are among the photosynthetic organisms that make up the base of the Antarctic food chain.

Oceanographer Mike Behrenfeld of the Environmental Protection Agency Oregon's office and other researchers believe that the 2 to 4 percent annual loss in phytoplankton is unlikely to worsen global warming. They estimate that decreases in CO_2 uptake due to the drop in productivity are trivial compared to uptake levels worldwide.

Nevertheless, researchers predict that increased UV will have damaging effects. Springtime Antarctic waters are cradle to many hatching species. The larvae of some fish, like cod, float in the uppermost layer of the ocean. There they take a direct hit of UV radiation before their scales form. So scientists are broadening their concern to include fish populations. Researchers know that UV kills fish larvae, but so do many other things — the life of a larva is so precarious that only one in a million lives to be an adult. So more studies are needed to isolate the impact of the springtime UV rise on fish larvae.

But even if the extra UV effects on the fish larvae prove to be as minor as the effects on the phytoplankton, Antarctic life won't get off scot-free. That's because the plants and animals of the southern waters

JANUARY 1993 **17**

are not merely a food factory or a system for converting solar energy to biomass. They also make up a community of many different species that depend on and compete with each other. Increases in UV may change not only the total productivity of Antarctic life, but also the apportionment of that productivity among the different species. That's because some species may adapt better to the increased UV and expand in number at the expense of other species. The results of such a change are difficult to predict, particularly since small annual effects can add up over the course of many years. This is true not only of phytoplankton in the Antarctic, but also of living things all around the world, including both wild and agricultural plant and animal species.

Not all plants and animals respond to UV with the same effectiveness. Some species work at night to repair damage to their DNA from exposure to ultraviolet radiation during the day. Thus, their protective mechanisms are retroactive. Other species protect themselves proactively with chemical shields analogous to the melanin in human skin. These organisms alter their chemistry in response to rising levels of solar radiation in the spring so they can dissipate the damaging energy of the UV radiation. (Indeed, an Australian company is trying to harness a phytoplankton UV blocker for use in sunscreens.)

In many species, the buildup of visible light that begins in springtime triggers proactive protection. But when the ozone hole opens in spring, the most damaging wavelengths of ultraviolet radiation, called UV-B radiation, begin to shine intensely before visible light is strong. So species with protective mechanisms triggered by visible light may be more vulnerable than species with other protection.

Within a community of species like phytoplankton, productivity of vulnerable species may fall precipitously, leaving room for better-protected species to grow strongly. The net result may be little or no change in the community's overall productivity, yet the makeup of the community has changed substantially.

"The point is that fish larvae and phytoplankton have already figured out how to deal with high UV or there wouldn't be anything in the tropics," where UV levels are naturally high, says Russ Vetter, an ichthyologist at the National Marine Fisheries Service in La Jolla, California. "The real issue is — If an animal has adapted to something in its local environment and then you change it, will you kill it? And if you don't kill it, how fast will it adapt to new conditions?"

In the teeming communities of phytoplankton, zooplankton, and fish larvae that populate the surface of the southern ocean, the annual return of the ozone hole and its UV beacon is likely to change the relative abundance of species, causing disruptive effects to ripple all the way up the food chain.

Ray Smith's group looked at just such species selection on their trip to the Antarctic. Comparing two individual species of phytoplankton, they found that UV stunted the growth of one predominant species far more than another. Diatoms, which look like spiny boxes, suffered far less than *Phaeocystis*, which grows in colonies encased in a jelly-like material, says marine biologist John Cullen of the Bigelow Laboratory for Ocean Sciences in West Boothbay Harbor, Maine. "There's a big difference in the kinds of ecosystems you can support if *Phaeocystis* dominated as opposed to if this diatom dominated," Cullen says.

Their feathers will shield penguins from most of the direct effects of increased UV, but changes in the food chain could lower supplies of their favorite foods and affect their populations.

As the ozone layer thins over the middle and lower latitudes, exposing forests, grasslands and cultivated croplands to larger and larger doses of UV, some plant researchers believe the same sort of community shifts will occur. These scientists have looked at the effects of UV on about 300 plant species and varieties within species, including cultivars of wheat, rice, soybeans and corn — the major agricultural crops. They've included a smattering of cucumbers, petunias, sunflowers and loblolly pines to round out the picture. Given an extra dose of UV (from sunlamps) the plants' responses run the gamut. Between a half and two-thirds of the species studied showed significant reductions in growth, according to Joseph Sullivan, a botanist at the University of Maryland in College Park. "Some are not affected as much as others and some don't seem to be bothered at all by the extra light," he says.

Considering this wide variability, Sullivan says, "I don't think you'll see huge losses in productivity on a total ecosystem basis because there are plants out there that can do well. I think what you'll see are changes in competitive balance, in community composition, and potentially in which crops or which varieties of a crop are grown in a certain area."

In the wild, changes in the makeup of the plant community will, in turn, affect the animals that can thrive there. "In a natural ecosystem the plants and animals have all evolved together," says University of Maryland botanist Alan Teramura. "And when you start changing the composition of the plant community, the animal community composition will be forced to change as well."

Such a dramatic shift in the plant community is likely to be lethal only for creatures that are very particular about the food they eat — such as the Chinese panda bear, which dines almost exclusively on a few species of bamboo. If bamboo happens to be particularly vulnerable to UV, then the panda would be wiped out. "I suppose extinction is a fairly radical thing that might happen," says Sullivan, "but it's not beyond possibility with extreme ozone depletion."

Most researchers believe that both land and sea ecosystems will eventually adapt to the increased UV but in ways that are not yet understood. One example of the uncertainty is a handful of reports that UV can compound the effects of other environmental stresses on plants, such as pollutants in the soil. A recent experiment on Norway spruce seedlings showed a 33 percent drop in photosynthesis when the plants were exposed to both UV and the toxic metal cadmium instead of either alone. Similarly, a higher dose of UV can make plants more susceptible to disease or make the effect of a disease more severe. Scientists are only beginning to explore such interactions.

The impact of increased UV levels on the global ecosystem may also create feedback mechanisms that could worsen the environmental problems.

Researchers are studying the effects of increased UV radiation on terrestrial plants. Here, student Dikang Wan examines loblolly pines growing under sun lamps.

Several researchers note that phytoplankton produce a kind of natural antifreeze called dimethyl sulfide, or DMS. DMS protects phytoplankton from freezing in icy Antarctic waters. When the phytoplankton die, some of the DMS in their cells is released into the atmosphere. Once there, it acts as a nucleus for cloud condensation. If increases in UV radiation force a shift in the phytoplankton community toward a variety that makes less DMS, there could be a decrease in clouds. Clouds help screen out UV, so fewer clouds could push UV levels even higher, plunging the climate into a spiral of dwindling clouds and mounting UV.

Researchers don't know if this scenario is likely. Nor, for that matter, can they identify the many possible effects of UV on climate or on life. Yet changes known and unknown are underway. For decades to come researchers must play scientific catch-up to figure out how and where damage has been done and will be done. ⊕

Shawna Vogel was a 1992 Knight Fellow in science writing at the Massachusetts Institute of Technology. She is writing a book on the dynamics of the Earth's interior.

OIL
When will we run out?

Imagine the Great Lakes drained of water. Then imagine pouring into the lakebeds all the oil we know of or anticipate ever finding and extracting. A small puddle in Lake Superior's basin, less than 5 percent of the lake's volume, would represent the world's entire oil endowment, for all time, for all humanity.

Wars have been fought over the control of oil. Wars of words have been waged over the rights to drill for it. And oil itself has been a weapon of war. Yet schoolchildren and heads of state alike often fail

by David G. Howell, Kenneth J. Bird and Donald L. Gautier

The first oil well was drilled in 1859 near Titusville, Pennsylvania. By 1990, there were 597,320 producing wells in the United States.

David Howell

Oil's distribution and abundance are controlled not so much by the properties of the substance itself as by the way it moves through the Earth's crust. Still, it is helpful to begin with a definition. Oil is a substance, usually liquid, made up of chains of carbon and hydrogen atoms. It forms in nature from the breakdown of organic molecules most people know from food labels: fatty acids, carbohydrates, sugars, proteins. Any living thing can provide these materials for the formation of oil, but phytoplankton — single-celled marine and aquatic plants — is by far the most abundant source.

For oil to form, phytoplankton must be buried beneath thick layers of rock. Most phytoplankton never turns into oil because it is eaten or decomposed by other organisms before it can be buried deeply. But in the past, oxygen-depleted waters sometimes bathed broad inland seas and continental shelves. Phytoplankton-eating organisms could not survive in these environments, so organic material could accumulate on the seafloor. At times the proportion of organic material reached 30 percent or more in some mud rocks, compared to 1 percent or less for most present-day sea and lake sediments.

Transforming these organic compounds to oil requires heat. Molecules of fatty acid and the like are robust and can remain unaltered in rock for millions of years. But the heat of the Earth can excite their atoms and split their chemical bonds, allowing oil to form. The temperature of the outer part of the Earth's crust rises approximately 2° F with every 100 feet of depth. Two miles below the surface, the temperature is sufficient to begin transforming organic chemicals from phytoplankton into oil. But not far below that the temperature gets too high, and the oil molecules themselves begin to break apart.

It is not possible, however, to find oil reserves simply by looking for places where organically rich sediments are buried two miles deep. Oil initially forms small dispersed droplets that we cannot recover economically. These droplets must accumulate in large quantities before the oil becomes useful.

As oil forms within rock and pressures gradually increase, it is squeezed right out of the rock. At depths where oil can form there are no large holes or tunnels in the rock through which oil can move. Instead, oil droplets seep through a network of microscopic pores and fractures. The larger the openings, the easier it is for the oil to migrate, but the rate of movement is always painstakingly slow, measured in inches per year.

Oil is lighter than either rock or any water that may be present, so it will rise buoyantly toward the Earth's surface or move laterally in the direction of lower pressure unless it is trapped by a layer of impenetrable rock. If the layer below this cap is itself highly porous, it can act like a sponge and soak up the oil. Only when oil enters such a capped geologic structure does it become a resource useful to human endeavor. Subterranean rocks in many different con-

to understand the principles that determine how much oil there is and where it is deposited. Many people have been lulled into complacency because oil has been so cheap most of this century. Supplies seemed boundless as discoveries outpaced consumption. But since the mid-1970s new discoveries have lagged and the limits of the oil endowment have become more obvious. Suddenly the oil finders began striking out, despite worldwide sleuthing and an increased understanding of oil's geologic habitats.

By the late 1950s, when geologists first proposed plate tectonics, we had found probably half the oil we'll ever discover. It was easy to find. But today oil exploration focuses on deposits lying in less-obvious places and finding them requires greater effort. So we have begun applying insights from the theory of plate tectonics, which states that the 30- to 90-mile-thick outer layer of the Earth is divided into slowly moving plates. Twenty years ago, the three of us joined a research team at the U.S. Geological Survey interested in using this theory to find and estimate the amount of oil left to be found.

Geologic understanding doesn't just tell us where oil is; it also tells us where it isn't. In recent years a synthesis has gradually emerged, linking our understanding of oil formation and movement with plate tectonics to enable us to plot the distribution of all the world's oil reserves — even in regions that have not been explored for oil. Thus petroleum geologists can make reliable estimates of our total oil endowment. The result is not a warning of immediate doom. It is, however, a clear message that the time has come to begin planning for a post-oil world so that when our oil supply runs out, as it could in the next century, we will have other energy sources ready.

figurations can trap oil, but worldwide almost all oil has been found in arched or dome-shaped traps, called anticlines in geologic parlance.

Generations of oil geologists have used surface geological mapping and seismic soundings to search for anticlines and thus for oil. But plate tectonics now helps explain how these anticlines are distributed. The Earth's plates move with the speed of growing fingernails, but their effects are powerful enough to generate great earthquakes, volcanoes and chains of mountains. They are also powerful enough to build the structures that contain the world's valuable oil pools.

Domes and anticlines most often occur where tectonic forces squeeze the Earth's crust, in regions where continents have collided or oceanic crust is moving toward a continental margin. The force of these movements wrinkles the crust just as one can wrinkle a rug by pushing its sides together. Anticlines also form where continents are being stretched apart. When horizontal layers of rock are pulled along a diagonal fault, some of the layers may lose their support and collapse into an archlike form.

Most accumulations of oil, at least 70 percent, are associated with areas where plates converge. The huge reserves of oil in the Middle East lie near the collision zone between the Arabian and Eurasian plates. The oil north of the Brooks Range of Alaska and east of the Urals in Russia results

Oil is not commercially exploitable until it pools in large quantities. Usually this happens in geologic formations called anticlines, rock strata that take the form of arches or domes. Calico Bluff, along the Yukon River in Alaska (above), exposes an anticline in unusually vivid detail. Below: If one of the layers in an anticline is impermeable, oil will be trapped beneath it, forming a pool in the permeable levels below.

MARCH 1993

David Howell

from the convergence of crustal plates. This knowledge of plate tectonics enables us to predict that we may discover new supplies of oil in the continent-side foothills of the Southern Andes and the interior basins of China. (See "Xinjiang, China: Between Two Mountains," *Earth*, September 1992.)

Most of the remaining 30 percent of oil lies in areas where plates have rifted, spread apart from one another. The oil fields along the margins of the Atlantic Ocean off Brazil, off Nigeria, and between Britain and Norway in the North Sea, or in the Mediterranean in offshore Libya all result from continental rifting. Geologists predicted that Georges Bank, off the New England coast, would also produce oil because of its rift-derived structures. Little oil has been found there, however.

Anticlines and other suitable habitats for oil may also form where plates slide past one another, as they do along the San Andreas Fault in California. But these occurrences are rare. The oil in the eastern part of China is astride a similar ancient plate rupture.

A brief history of plate movements and oil formation in the United States will show how tight the relation is between the two. Approximately 300 million years ago the continents of the Earth began colliding to form the ancient supercontinent of Pangaea. The forerunners of Africa and South America smashed into the eastern and southern margins of North America's precursor. These collisions produced the settings for oil and gas we are now exploiting along the landward margins of several mountain ranges, including the Marathon Range in Texas, the Ouachita Range in Arkansas and Oklahoma, and the Appalachians. About 160 million years ago South America rifted away from North America, bringing into being the Gulf of Mexico. The gulf basin contains numerous oil-producing anticlines (hence the name Gulf Oil). Between 80 million and 40 million years ago tectonic forces resulting from North America's westward drift forced up the Rocky Mountains. Oil basins formed at that time in Colorado, Wyoming and Utah. Finally, 20 million years ago, the Pacific Plate began sliding along the western margin of the United States. The oil habitats of California formed at that time. Thus, we can see that in the United States alone it has taken over 300 million years of tectonic activity to produce a supply of oil we will likely consume in just 200 years.

The world's last oil discovery has surely not been made. But we are just as surely consuming oil at a rate that vastly exceeds the rate at which it is forming and accumulating. Our oil supply formed in units of geologic time — millennia, epochs, eons — but it is being used in units of human time — centuries, decades, years. Ever since Western economies have depended upon petroleum for power and transportation, people have wondered how much oil exists. The future of civilization turns on the answer to that question.

The same geological understanding that helps us search for oil can help us find that answer. Though we cannot predict how much oil we will find in a given zone of compression, rifting or strike-slip movement, we can use our knowledge of the tectonic history of the Earth to identify those areas whose oil deposits we must consider in making a global estimate. Then we can turn to other techniques to estimate the amount to be found in these areas and produce a total figure and feel confident that we have not overlooked any major parts of the world's oil endowment.

Even with the help of plate tectonic theory, it would be impossible to estimate with any certainty the absolute quantity of oil currently existing on Earth. But much more important than the overall abundance of oil are the size and number of oil accumulations that can be exploited for purposes of human activity. Oil occurs in discrete accumulations that range in size from the tremendous Ghawar field in eastern Saudi Arabia (the largest oil field in the world with ultimately recoverable reserves probably in excess of 100 billion barrels) to the innumerable tiny droplets of dispersed oil found in many sedimentary rocks. There are many more small deposits than large ones, yet most of the world's oil is in the few large accumulations. There is a surprising number of very small fields, but they do not make an important contribution to the total supply. In addition, they rarely provide a financial return that justifies the cost and energy required for discovery and extraction. Therefore we can consider only moderate and large deposits in estimating the world's total volume of recoverable oil.

There are three components to the world's oil supply. The first is production, the oil that has been and is being produced. The second is reserves, the oil

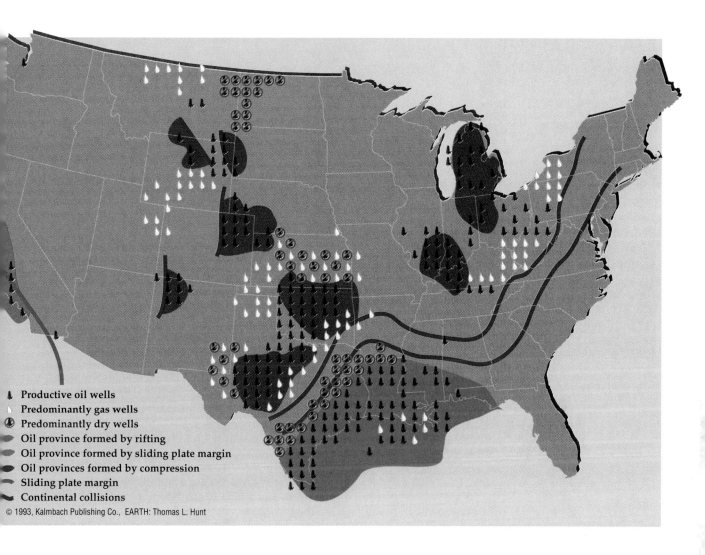

- Productive oil wells
- Predominantly gas wells
- Predominantly dry wells
- Oil province formed by rifting
- Oil province formed by sliding plate margin
- Oil provinces formed by compression
- Sliding plate margin
- Continental collisions

© 1993, Kalmbach Publishing Co., EARTH: Thomas L. Hunt

that has been demonstrated to be available for production and that can be delivered efficiently through the existing infrastructure of wells. The third is the oil that remains to be discovered, usually referred to as undiscovered resources.

One of these components, reserves, is more slippery than it might initially appear. The world's estimated oil reserves now stand at roughly one trillion barrels. But experience in the extensively developed oil fields of the United States has shown that reserve numbers change through time. For example, in the United States, even fields discovered fifty, sixty, or seventy years ago continue to display reserve growth. This growth of reserves occurs for a variety of reasons. Sometimes the initial estimates were overly conservative. Sometimes we discover new pools or parts of pools in known fields. And improved technology may increase the efficiency of oil recovery. Growth of known fields is now almost as important as discoveries of new fields in maintaining the reserves of the United States. This startling fact may eventually have worldwide implications, but at present reserve growth is largely a phenomenon of the United States.

Pulling together estimates made by all methods for all of the oil fields and potential oil fields in the world, we find the estimated global oil balance. Our estimates turn out to be near the middle of estimates

The productive oil wells of the United States all occur in areas influenced by three kinds of tectonic movement. West Texas oil country, for example, results from an ancient continental collision, the Gulf of Mexico has been the scene of pull-apart, and California has been shaped by shearing motion.

made by other geologists. Through 1990 the world has produced and consumed 650 billion barrels of oil. Another 950 billion barrels of recoverable oil are known to exist in discovered fields. (Only about 30 percent of the in-place oil is recoverable. The rest remains bound to rock particles by surface tension and other forces.) Estimates of undiscovered resources range between 300 and 900 billion barrels, with a most probable mean estimate of 500 billion barrels. The following table summarizes these figures, in billions of barrels of oil.

	U.S.	World
Production (through 1990)	155	650
Reserves (1991)	25	~950
Undiscovered recoverable oil	~40	~500
Anticipated reserve growth	~20	Unknown
Total recoverable oil	~240	~2,100
Total oil left for the future	~85	~1,450

Thus, the total of what has already been used, what we know exists, and the best estimate for what is yet to be found is roughly 2 trillion barrels of recoverable oil. These figures are not as large as they seem. If all 2 trillion barrels were spread at one time evenly across the continental United States, they would form a layer less than 2 inches thick. If this layer were divided among all of Earth's 5 billion people, each would be allotted a portion slightly smaller than half a football field.

All of the oil that has been discussed thus far comes from what are considered conventional oil resources, those that can be easily extracted by drilling. Besides conventional oil, we know of enormous accumulations of unconventional oils, such as the extra-heavy oil in the Orinoco oil belt of Venezuela, the Athabasca tar sands of western Canada, or the cold, viscous oil accumulations of the North Slope of Alaska (yet to be extracted). Shale oil also has a resource potential, although strictly speaking it is not yet oil. Rather, shale oil is a source rock, an organic-rich sedimentary rock that has not yet been "cooked" enough to transform the organic material to petroleum. We can produce oil by heating the shale in an oven, but cost estimates, not including environmental costs, run at least 2 or 3 times the present-day cost of conventional oil.

We mention these unconventional resources because they will become increasingly important in the future. But for all of them the cost of exploitation is high and their ultimate resource potential remains clouded in both economic and geologic uncertainty. It is unlikely that we will ever use large amounts of these unconventional oil supplies, even though the

The world's remaining major oil discoveries will probably come from zones of compression, pull-apart and shear — the same kinds of regions where oil has been found in the United States.

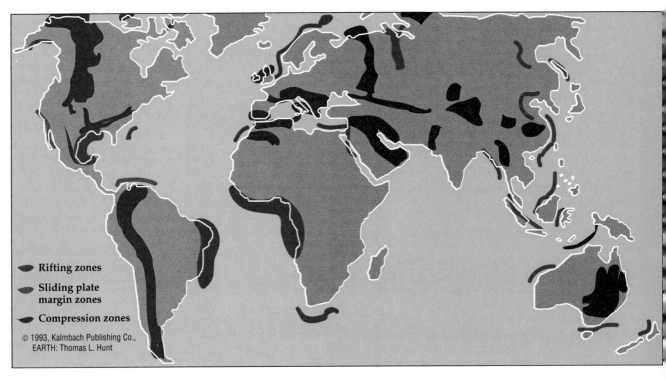

- Rifting zones
- Sliding plate margin zones
- Compression zones

© 1993, Kalmbach Publishing Co.,
EARTH: Thomas L. Hunt

volume of oil in these deposits exceeds the original volume of conventional recoverable oil.

All indicators suggest that the United States is running out of oil. The fields we are discovering are smaller. The amount of drilling necessary to discover a barrel of oil is increasing. The technology needed for recovering the hard-to-get-at oil and gas that is left is becoming expensive. Since the 1960s, the United States has consumed more oil than it produces, and the shortfall has grown over the past three decades in spite of unprecedented levels of drilling. By the late 1980s the United States was importing more oil than it was producing.

Dreary as the domestic situation seems, the world oil supply provides a temporary comfort zone. Reserves in the Middle East plus anticipated discoveries there amount to about 750 billion barrels. Elsewhere in the world are another 650 billion barrels or so. At the present rate of worldwide consumption of 20 billion barrels per year, this would suggest a 70-year supply. (It is likely that consumption will go up as developing countries attain higher standards of living.) But what will happen — politically, socially and economically — as we try to distribute this oil? What will be the pressures as we approach the final drop of consumption?

We've got enough oil left to plan prudently for the future. As oil prices increase, we know that some unconventional supplies of oil will become economically viable. In addition, energy gases, along with solar power, wind and geothermal energy could provide a comfortable transition, a bridge to the energy supplies of the twenty-second century when perhaps nuclear fusion will provide energy for society.

There is also coal, with its well-recognized abundance and accessibility. But coal is a worse polluter than oil or gas. Scrubbing technologies, if properly implemented and maintained, reduce sulfuric acid pollution, but the emission of greenhouse gases (carbon dioxide and methane) is and will continue to be worrisomely high. We hope that humanity will make the transition without increasing its dependence on coal. But if demands for energy increase as rapidly as the growth in world population, then all bets are off. The danger is that the lights too may go off.

A WORLD WITHOUT OIL

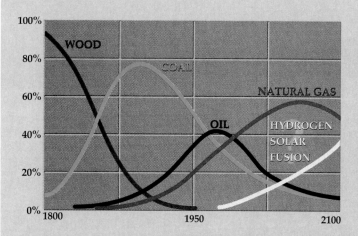

Nebojsa Nakicenovic and Cesare Marchetti of the International Institute of Applied Systems Analysis plotted past use of energy from various sources and projected future use.

We may now be near the peak of our reliance on oil. Other energy sources are already beginning to take its place. Natural gas consumption is rising and may peak between 2030 and 2050, providing us with 50 percent of our energy. Coal consumption is rising too but it is uncertain how long we can continue to use coal, in view of the pollution and carbon dioxide released when it burns. There are also several nonfossil energy sources that may take us into the twenty-second century.

- Solar and wind: These clean sources have yet to reach levels of efficiency that would enable them to compete with the fossil fuels. Increased research, combined with the rising cost of other energy sources, could change that soon.
- Fission: Nuclear energy from fission has provided a steady 5 percent of the world's energy requirements since the 1960s. Technological changes could raise that figure by making it possible to operate plants at the level of safety that the public demands. But no one has yet found a way to handle nuclear waste to the public's satisfaction.
- Fusion: Controlled sustained nuclear fusion remains an immense unsolved engineering problem, but fusion uses cheap abundant materials as fuel and would produce much less radioactive waste than fission. The date when the problem is projected to be solved has been pushed back again and again and now stands at 2020 or later.
- Hydrogen: Whatever source our future energy comes from, we will want to be able to move energy from place to place. Oil derivatives like gasoline now enable us to do that with ease. In the future, energy derived from the Sun or fusion may be used to extract hydrogen from water. That hydrogen could then be carried in a tank and burned for fuel, producing only water as a waste product.
- Efficiency: It is almost always cheaper to lower energy demand, by using more efficient appliances, for example, than it is to increase supply. — *Tom Waters*

David G. Howell, Kenneth J. Bird and Donald L. Gautier work in the Division of Petroleum Geology of the U.S. Geological Survey, Menlo Park, California.

POWDER RIVER COAL:
Geologic enigma, environmental dilemma

by Doug McInnis

A coal truck rumbles through an AMAX open pit mine near Gillette, Wyoming, in the Powder River Basin. This region, which straddles parts of Wyoming and Montana, is a major producer of low-sulfur coal and a locus of both scientific interest and environmental controversy. Geologists are puzzled by the basin's mysteriously thick seams of coal. Environmentalists are concerned about the ecological impact of strip mining the coal.

The Powder River Basin is 9,700 square miles of northeast Wyoming and southeast Montana, wedged between the Black Hills of South Dakota on the east and the Big Horn Mountains on the west. In the last century it was known for outlaw Butch Cassidy, who drank in the bars of the Powder River Basin town of Buffalo and hid in the nearby Hole-in-the-Wall country from which his gang drew its name. In the 1920s the basin was the site of the Teapot Dome oil field, which brought scandal to Warren Harding's administration.

But in the last quarter of this century, the region has gained fame for its coal. The Powder River Basin is believed to contain 56 billion tons of low-sulfur coal — enough to power the nation at current consumption rates for about 50 years. Coal companies value the basin because it contains huge amounts of coal right near the surface, enabling them to strip mine it inexpensively. Geologists value the basin because the coal presents a marvelous geologic mystery.

Yet Powder River coal poses an environmental problem. On the one hand, it's low in sulfur, which means it is less polluting than coal from other regions. Strip mining, on the other hand, forever alters the fragile landscape. The mining companies say they are restoring the land to health and productivity. But environmentalists say that despite these efforts, strip mining causes pollution and loss of habitat for native plants and animals.

Coal from a thick seam is loaded onto a truck for transport out of the AMAX Eagle Butte Mine near Gillette. Water is sprayed to prevent coal dust from blowing into the air.

Powder River's coal seams run remarkably thick and unsullied by other material. Usually, unwanted sediment such as clay washes over a deposit before coal seams can get very thick. In the eastern United States, for instance, coal seams tend to run five to ten feet thick and can be much thinner. But Powder River coal is packed in immense strips, some more than 200 feet thick. These seams stretch vast distances up and down the basin. "They're hundreds of miles long. They're 50 miles wide," says James McClurg, a geologist at the University of Wyoming. "They're not little pods of an acre or two. They're immense things." McClurg, who has studied the basin for more than a decade, says no other place in the world has as many seams 50 feet or more thick.

But in spite of the coal's abundance and accessibility, large-scale mining didn't take hold in the basin until more than a century after early explorers and surveyors noted its rich deposits. Development was first held back by a treaty with Plains Indian tribes that placed the land off limits to development. Then gold was discovered in the nearby Black Hills. Prospectors swarmed through the basin, violating the treaty and exacerbating tensions with the tribes. In the midst of all this, General George Armstrong Custer led his troops to disaster at the Little Big Horn in 1876. The government retaliated, repudiated the treaty and eventually forced the Indians from the region.

During the next 100 years oil, natural gas and uranium began to flow from the basin. But still not coal. Powder River coal had two drawbacks. First, it was too far from big-city markets to justify the freight cost. Second, it had a lower heat value than its eastern cousins. Coal from Kentucky or West Virginia typically has a heat content of 11,000 or more British thermal units per pound. Powder River coal's heat content is usually about 8,600 Btu per pound.

But during the environmental movement of the 1970s, the chemistry of Powder River coal became one of its prime selling points. Though the basin's coal packs less heat per pound than eastern coal, it is considerably lower in sulfur. Eastern coal often has a sulfur content of four or five percent, while Powder River coal is about one-half of one percent sulfur. This difference became critical once environmentalists focused national attention on acid rain, linked to damage of lake ecosystems and possibly to the death of trees. Sulfur dioxide and nitrogen compounds, which cause rain and snow to acidify, are released when coal and other fossil fuels are burned.

The Clean Air Act of 1977 required utilities to cut their sulfur dioxide emissions. To comply, the companies had a few options. Either they could install costly equipment to remove sulfur dioxide before it spewed from the stacks, or they could burn coal so low in sulfur that the equipment would be unnecessary. Suddenly, mining the Powder River Basin seemed very attractive.

The act gave low-sulfur western coals a competitive edge they'd never enjoyed before, even though Wyoming and Montana combined contain nearly 40 percent of the nation's coal reserves. By contrast West Virginia and Kentucky, whose names are virtually synonomous with coal, contain only 15 percent. Today the seven largest coal mines in America lie in northeast Wyoming and southeast Montana. Wyoming, once a minor coal state, has now emerged as the nation's largest producer.

AMAX, a major minerals company, opened the first mine in the Powder River Basin. Exxon, Mobil, Shell, Sun, ARCO and other energy companies soon followed suit. The companies brought tax revenues to a depressed western economy. They were drawn to the basin because it is ideal for strip mining on a gargantuan scale, making extraction efficient and cost-effective. In the eastern United States, relatively thin coal seams often lie deep underground, requiring tunnelling technology

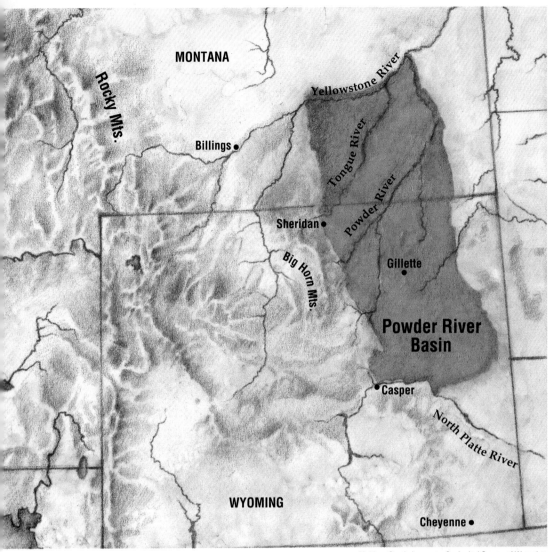

Matthew Groshek; source, Geological Survey of Wyoming

and many more workers to extract the coal.

In 1991, it took 30,000 Kentucky miners to dig 174 million tons of coal. That same year, Wyoming needed only 3,300 miners to remove 194 million tons. Giant machines stripped a thin layer of overlying rock away and then crews blasted the seams into manageable bits and scooped up the coal.

Naturally, this ease of extraction has made the coal relatively cheap. Wyoming coal often sells for $5 a ton on the spot market. Eastern coal, on the other hand, frequently costs more than $20 per ton.

But the Powder River Basin is not only an economic resource. To geologists, it's also an intriguing scientific enigma. Geologists have been studying the basin for more than a century, largely to answer a baffling question: How did the seams get so massive? Or more precisely, why weren't the seams diluted by influxes of clay and other impurities before they thickened? Researchers estimate that the coal must have accumulated for more than 100,000 years without being contaminated by impurities. The coal deposits thickened at a glacial pace — about one millimeter a year. As McClurg puts it: "How on Earth did things remain constant enough so that you get 200 feet of coal?" He and others have posed a few theories.

Geologists agree that coal forms in wetlands. Powder River coals formed some time during the late Paleocene and early Eocene, some 40 million to 50 million years ago, when the now arid grasslands brimmed with dense, temperate-to-subtropical forests of *Glyptostrobus* (related to the cypress), and broad-leaved cycad-like plants, as well as dawn redwoods, gingkos and other species. The coals formed when plant matter was transformed into peat, which was then buried under thousands of feet of sediment. The sediment provided pressure and blanketing insulation, allowing the Earth's internal heat to cook the peat and convert it to coal. After the coal formed, this blanket of sediment slowly wore away, ultimately leaving the seams close to the surface.

But researchers part company when explaining the precise conditions that created the Powder River coals. A widely accepted theory comes from Romeo Flores, a geologist with the U.S. Geological Survey in Denver. Flores concluded that Powder River coals evolved in a bog or swamp that was raised above the surrounding river landscape, not unlike the raised bog systems now found on the lush island of Borneo. Sediments brought in through river systems usually sully peat deposits in bogs, but raised bogs are protected from these influxes because they lie above the rivers. Over a long time, peat deposits can be transformed into very thick clean coals, like those found in the Powder River Basin. Flores theorizes that the basin bogs were sustained by drenching rains, as are the raised bogs in Borneo today.

McClurg casts the scene differently. He contends that the Powder River Basin contained a very large swamp that was fed and drained by rivers. The rivers brought in

30 EARTH

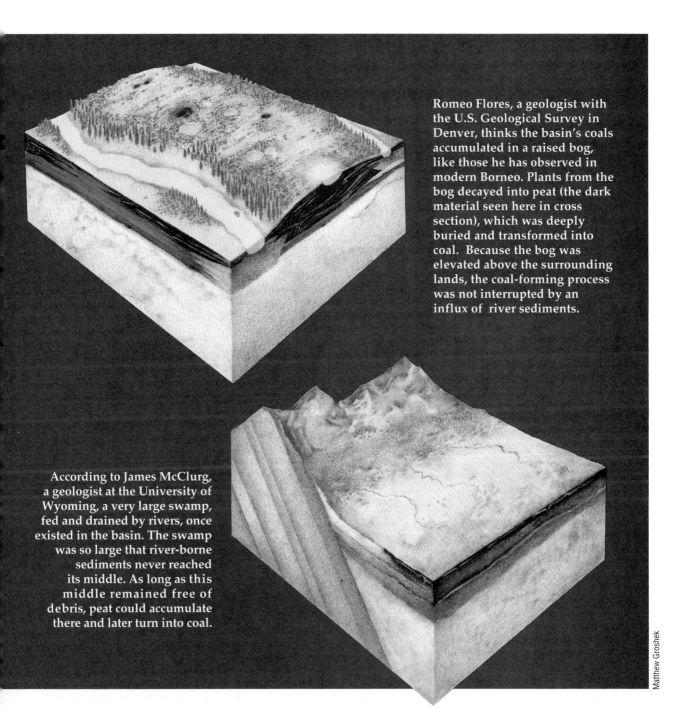

Romeo Flores, a geologist with the U.S. Geological Survey in Denver, thinks the basin's coals accumulated in a raised bog, like those he has observed in modern Borneo. Plants from the bog decayed into peat (the dark material seen here in cross section), which was deeply buried and transformed into coal. Because the bog was elevated above the surrounding lands, the coal-forming process was not interrupted by an influx of river sediments.

According to James McClurg, a geologist at the University of Wyoming, a very large swamp, fed and drained by rivers, once existed in the basin. The swamp was so large that river-borne sediments never reached its middle. As long as this middle remained free of debris, peat could accumulate there and later turn into coal.

sediment that collected around its edges but never reached the middle. (This is precisely what is happening today in the Okefenokee Swamp/Lake complex in Georgia.) Millions of years ago, the swamp may have dried up periodically or become so inundated with water that a lake formed. Both of these conditions would have halted coal formation. But as long as the river systems didn't carry debris into the center of the coal-forming region during the hiatuses between droughts and floods, coal could continue to form.

Walter B. Ayers and William R. Kaiser, geologists with the Bureau of Economic Geology at the University of Texas at Austin, crafted yet a third explanation. They argue that a delta region was created by rivers that brought in sediment from the Black Hills. Later, a swamp formed on the delta land. But by the time the coal began to accumulate, the river channels had switched course. That allowed thick coal seams to form, undiluted by influxes of river-borne clay or other materials. This theory doesn't preclude the possibility that raised swamps existed in the basin. However, the principal reason the coal-forming system avoided impurities was that the rivers changed course, the team concluded.

Powder River coal's low sulfur content is much easier to explain, geologists say. High sulfur coals form in areas exposed to ocean water, which contains sulfates. But McClurg says the oceans had drained away by the time the coal formed in the Powder River region during the late Paleocene and early Eocene. The scant amount of sulfur that is found in these coals was naturally present in the plants from which the coal formed.

MAY 1993 **31**

The genesis of any kind of coal demands a precise sequence of events, however. "If you don't get all of them, you don't get coal," explains McClurg. And to get seams more than 200 feet thick, low in impurities and low in sulfur, is a "remarkable achievement," he adds. "It would be like blindfolding yourself, spinning around and hitting the center of a dartboard 100 times in a row."

Environmental groups have recently taken as keen an interest in the Powder River Basin as geologists have. But they are concerned mainly about the impacts of large-scale strip mining on the region's delicate ecology. The National Wildlife Federation, the Sierra Club, and a smattering of state and local organizations have all spoken out about the strip mines, many of which lie on federal lands. Members of these groups believe the wholesale rearranging of the delicate western landscape of sagebrush and grasses can only spell trouble for the region's water supply, plants and animal populations.

Today state and federal laws require mining companies to reclaim the landscape following mining. This effort can be extensive and complex, involving replanting grasses, shrubs and other plants, recontouring to restore the flow of streams, and constructing rock piles to provide habitat for small mammals and nesting sites for birds.

But even the most conscientious restoration projects can't bring back the original landscape, mining companies concede. The question is: What is gained and what is lost? And how can these tradeoffs be measured?

Mining removes coal seams that serve as filters for underground water supplies. These filters aren't replaced. And when overburden is placed in the seams that once held coal, dangerous materials may come into contact with groundwater. "They can't really expect to reconstruct an aquifer," says Dave Alberswerth, public land specialist at the National Wildlife Federation. "Once you've got one of these big mines, you can change completely the surface geology." But coal companies stress that they are careful to keep heavy metals and other toxics away from groundwater supplies.

Removing overburden also often flattens the hilly rangeland that predated the mines. This alters patterns of vegetation and can decrease plant diversity, argues Tom Collins, environmental coordinator for the Wyoming Fish and Game Department.

Though mining companies replace the trees they fell, this activity itself has subtle ecological consequences: A newly planted cottonwood isn't as useful a nesting and perch site for hawks, eagles and owls as a 100-year-old tree, Collins says.

State and federal laws don't require companies to try to duplicate pre-mining plant communities, though. For instance, under Wyoming law, mining companies must try to restore just 10 percent of the sagebrush that once grew at a mining site, according to Collins. In his view, this is problematic. Sagebrush is a staple food for the pronghorn antelope that live in the basin. Since sagebrush regenerates only very slowly, it can take decades before plants grow tall enough to protrude above winter snows.

Also, mine lands are closed to hunters, which makes it difficult to control the size of pronghorn herds through hunting, state game officials say. The outcome could be too many pronghorn, too little winter forage and a large winter die-off through starvation. "Cumulatively the mining has taken out wildlife habitat that we think is significant," says Collins. "This is an ongoing thing. The reclamation is not going to be suitable as habitat for some years to come."

But mining companies have a different perspective. While mining inevitably changes the land, they argue that reclamation can actually enhance it. "[The land] comes back different, and in many cases it comes back better," argues Mickey Steward, environmental manager at AMAX. She and other

Mule deer roam the arid grasslands near a mine operated by the Antelope Coal Company in northern Converse County, Wyoming.

A Chicago & North Western train returns to the Powder River Basin to be reloaded with coal bound for midwestern markets.

An alert pronghorn stands in native shortgrass prairie in an undisturbed area of the Antelope Coal Company mine.

mining representatives say pronghorn herds are thriving and that the land is more productive now than it was before reclamation. "The area is flush with antelope," says Greg Schaefer, environmental manager at ARCO. "If we are impacting wildlife, how come we're not seeing it out there? It simply isn't happening."

Of course, there are environmental problems associated with coal burning as well as mining. The combustion of coal releases more greenhouse gases into the atmosphere than other fuels, a fact most climate scientists say will make a significant contribution to global warming and associated climatic upsets.

But the United States has not yet succeeded in weaning itself from coal. Although it poses many environmental problems, coal currently remains a vital fuel, so mines in the basin aren't likely to close down any time soon.

Some critics are learning to live with the Powder River Basin mines, hoping for the best. "We've never tried to stop the mining," says Dave Stueck, chairman of the Powder River Basin Resource Council, a local environmental group. "All we've said is that if there are problems, let's look for ways to mitigate them — to make sure that when the last mine closes its gate the people who choose to stay are not left high and dry." ⊕

Doug McInnis is a special projects reporter at the Columbus Dispatch *newspaper in Ohio.*

LIVE FROM SPACE

CLOUDS IN THE NEWS

by Steven A. Zaburunov

After a brief but noisy commute into low-earth orbit, the Columbia carries seven crew members of STS-40 toward a nine-day mission in space. (NASA photo S40 (S) 134 taken June 5, 1991.)

At some time in our lives, each of us has gazed toward the skies and wondered about the clouds. How did they get to be the way they are? Why are today's clouds completely different than yesterday's? Do you think that one really looks like Elvis?

Watching them move slowly against a brilliant blue background, we see only one small chapter in the life of a cloud, one piece of a larger puzzle. Shuttle crews have the unique advantage of studying cloud formations from a global perspective, watching patterns that stretch hundreds of miles.

These colorful images, all taken by astronauts on several of NASA's shuttle flights, allow us to share this global vision.

Some of the clouds on these pages are made by people, others by volcanos. Some are made by dusty desert winds, still others by the natural process of evaporation. A few are the terrible aftermath of war. But, as you will see, they are all easier to understand while studying their sunny side, circling Earth in the shuttle. The riddle of clouds unfolds when we share the astronauts' perspective — floating in space, daydreaming, locked in Earth orbit.

◄
This view is northeast toward Lake Poopo, Bolivia, and then over the Andes to lower elevations in Bolivia and Brazil. Extensive dry seasonal burning in the Amazon basin has produced a thick haze, which is trapped in the lower atmosphere by a stable upper layer. The Andes extend above the haze into clean air, preventing the haze from moving west. But just as the smoky air cannot cross over the high Andes, neither can the moisture-laden clouds, forcing them to release their moisture back into the Amazon basin, leaving the area in the foreground quite dry. The large white patches are playas, *ancient lakes that dried up, forming salt beds. The large playa is Salar de Uyuni in Bolivia. The photo was taken September 13, 1991, in Bolivia's mid-afternoon. (S48-151-139)*

►
Curious cloud patterns become apparent in the southern part of Florida. Much of the coast, including Cape Canaveral, is obscured by sea-breeze clouds. Note the absence of small cumulus clouds over bodies of water such as Lake Okeechobee. (For an astronaut's perspective, turn the magazine upside down.) (S40-613-049)

During the worst of the Kuwaiti oil well fires, an estimated five million barrels of crude oil burned per day, about three percent of total world consumption. As these photos show, nearly half were out by mid-September.

Although local carbon dioxide (CO_2) levels rose 15 percent, global levels have not been affected. Our atmosphere is so large that the additional CO_2 input was negligible at Hawaii's Mauna Loa measuring station.

The black clouds have lowered temperatures in their shadow some 5°C. Not only do the black clouds physically block the rays of the Sun, but they heat up, forming an atmospheric blanket that traps the cooler air below.

Most of the smoke particles usually settle out, except for the finest sizes, which might rise into the upper atmosphere and remain there for a long time. The smoke's direction changes day to day.

Sanjay Limaye, an atmospheric scientist for the University of Wisconsin-Madison, has studied the atmosphere of several planets in our solar system. He traveled to the Middle East to take a firsthand look at the smoky cloud. "I was in Bahrain, and the black cloud had touched down to earth. It was very smelly, and you could not see more than a mile. It was just like being in Los Angeles riding behind a bus."

Looking at the shuttle photos, you can see the improvements that were achieved from May to September. The northern fires of the main fields (south of Kuwait City) are almost entirely out

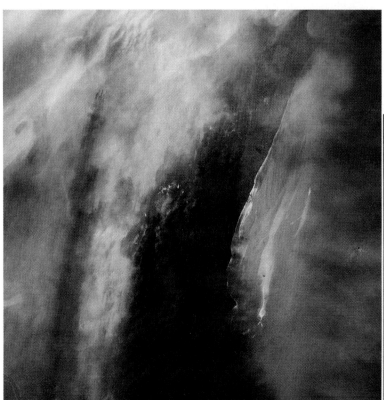

▲
The scene in early May of 1991, taken by the crew of the Space Shuttle Discovery (STS-39), was not pretty. Kuwait City is to the north; the wind drove the black clouds south the day this was taken. Oil slicks appear on the Gulf. The white clouds are dust storms in this infrared image. (S39-87-12)

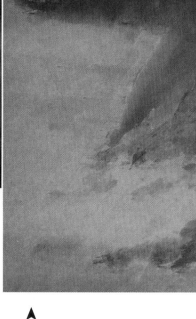

▲
Set afire by retreating Iraqi armed forces, oil wells to the north of the Bay of Kuwait and just south of Kuwait City, on the south shore, burn out of control. The crew of STS-39 also took this photo. (S39-72-060)

Viewed from orbit, clouds take on a whole new appearance.

Skies in the Middle East are normally free of clouds. The Sinai Peninsula dominates this north-looking view. The Red Sea (the large body of water in the foreground) is clear of river sediment because of the prevailing dry climate of the area. We see several rift systems, notably the great rift of the Gulf of Aqaba that extends northward to Turkey through the Dead Sea. The haze on the Mediterranean Sea is not clouds, but wind-borne dust. The international border between Egypt and Israel stands out clearly, reflecting different rural vegetation and irrigation practices. (S40-152-180)

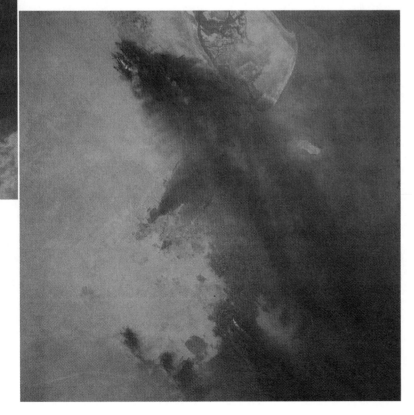

Crew members of the Space Shuttle Discovery (STS-48) took these shots of Kuwaiti oil fields in mid-September 1991. The number of oil well fires has been significantly reduced since May 1991. In this view, most of the remaining oil fires (about 300), are in the two largest fields, Sibirayah, north of Kuwait Bay, and the giant Magwas-Burgan-Al Ahmadi field, south of Kuwait City. (S48-605-030)

*T*he circular island is Gran Canaria of the Canary Islands, located in the North Atlantic Ocean just west of Africa (north is to the upper left). The wind is blowing from left to right in this angle.

Low-level stratus clouds have become trapped in vertical movement because of an overlying atmospheric temperature inversion. Because the clouds cannot flow over the island, they swirl around it. The wind, moving from north to south (or left to right), forms giant swirls downwind of the islands. These swirls, called von Karman vortices, can be traced for hundreds of miles.

The northern extent of a large dust storm moving off the coast of Africa is apparent in the upper right of the image. Later in the mission, the astronauts noticed that the dust cloud had blown across the entire Atlantic Ocean, appearing as far west as the Dominican Republic. Those on board the Space Shuttle Columbia (STS-40) made the observation in mid-June 1991. (S40-75-03)

◄

Rick Hauck, astronaut on the Discovery (STS-26), enjoys a view that every photographer would love to have. (S-26-47-13)

38 EARTH

Mt. Pinatubo is not the only volcano creating clouds that can be seen from space. These volcanos in central Java, Indonesia, are among the most active and explosive in the world.

Java has 35 active volcanos that form an east-west line of peaks down the spine of the island. Three of the five volcanos in this image are smoking. To the west, the largest plume is along the flank of Mt. Lawu. Actually, the smoke is not of volcanic origin, but comes from a large ground fire on the mountain's flank.

At Tengger caldera, on the eastern edge of the frame, a smaller plume is also from a ground fire, according to NASA. Only the plume flowing north from Welirang volcano (just east of the central cloud mass) is believed to be steam emissions from that volcano.

Kelut volcano, just south of Welirang, had a deadly eruption last year. The summit area is still ash-covered, and many of the river drainages on the mountain remain visibly choked with the resulting ash. The mid-September 1991 photo was taken from the Space Shuttle Discovery (STS-48). (S48-151-064)

*T*his wide-angle photo is centered over Kashmir, India, with a view north into the central Takla Makan Desert of eastern China (north is toward the right). The extensive layer of haze, visible from India to the West Siberian Plain, was much more pronounced than the Earth Observation Scientists at NASA had expected. This photo from the Space Shuttle Discovery (STS-48) was taken September 15, 1991. (S48-152-226)

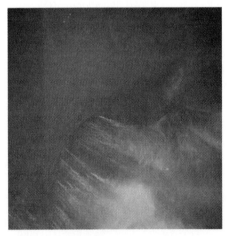

This area of Indonesian Borneo is known as Kalimantan. Fires and the associated smoke plumes indicate that the land is being prepared for new agriculture. (S48-71-90)

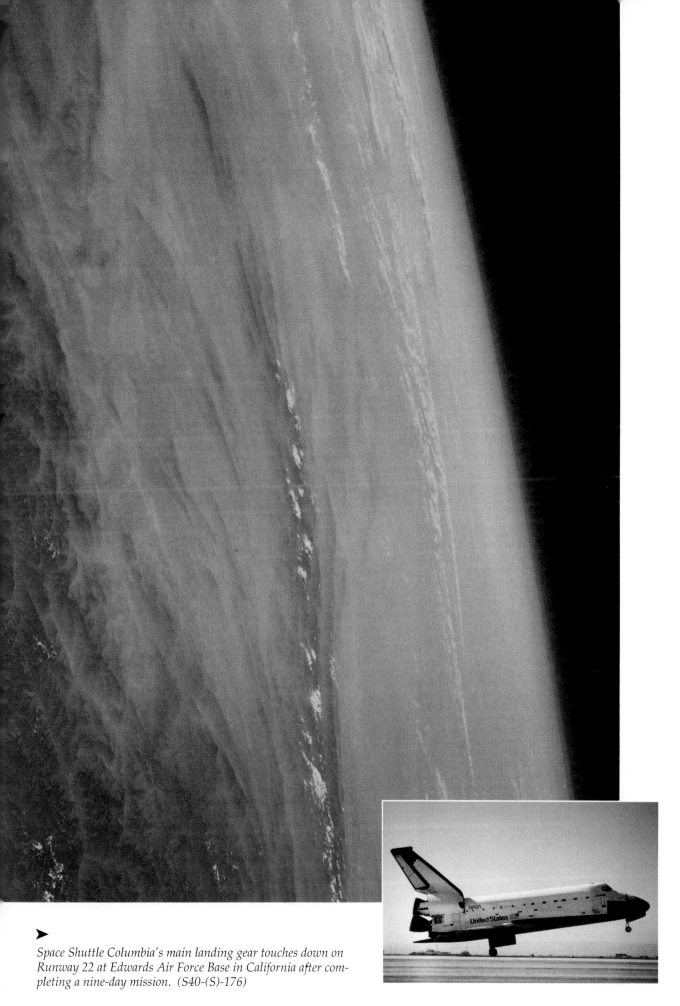

➤ Space Shuttle Columbia's main landing gear touches down on Runway 22 at Edwards Air Force Base in California after completing a nine-day mission. (S40-(S)-176)

In this GOES 7 image of the North Atlantic region, red and white indicate storm areas over blue water and green land.

Receive Satellite Images on Your Computer

By Philip J. Imbrogno

Satellites, those amazing inventions of the twentieth century, provide us with an opportunity to study our earth from space. From a viewpoint hundreds of miles above the earth's surface, we study and observe many forces of nature such as hurricanes, volcanic eruptions, droughts, floods, and even the aftermath of earthquakes.

Ten years ago you needed to be a professional to receive satellite pictures of Earth. Today, because of the boom in communications and computer technology, you can set up your own earth-based receiving station with store-bought amateur radio equipment, a simple antenna, some software, your computer, and this article. Although the price isn't exactly in the pocket-change category, you should be up and running for under $1,000 if you already own a computer.

The images you can receive are from weather observation satellites such as those launched by the United States and the former Soviet Union. Over 2,000 satellites now orbit the earth. Some actively send data; some do not. If you know their orbits and their operating frequencies, you can pick up a number of satellites with your receiver.

Many satellites broadcast at frequencies close to the 2-meter amateur radio band, so you can take advantage of a wide selection of available equipment. [If you are unfamiliar with some of the equipment or computer terms, a knowledgeable salesperson can often provide valuable assistance.] Antennas are no problem because you can easily build or modify the antenna you need. With little effort and time invested, you will be able to receive and image American, Russian, Chinese, Japanese, and European satellites.

The satellites you will be able to image from your receiving station typically have polar orbits or remain fixed in the sky over a single location. Polar-orbiting satellites scan the entire earth from north to south, or from south to north, from a height of about 300 kilometers (200 miles), making two passes per day over a single region. Most satellites are placed in orbit in this fashion, as each orbit allows them to pass over a different area of the earth. The fixed satellites (called geosynchronous or geostationary) view an entire hemisphere from their stations about 36,000 km (22,000 miles) above the equator. (See diagram on page 59.)

The meteorological polar-orbiting satellites operated by the U.S. Weather Bureau image (or send computer data used to create a picture) the cloud cover over North America and the rest of the world. Because these polar-orbiting satellites cover the entire planet, they operate over a 24-hour period and are used quite frequently.

These satellites image the earth by a technique called Automatic Picture Transmission (APT). This technique, similar to a telephone fax transmission, allows scientists (and you) to receive real-time images of the earth whenever the satellite comes within range of the receiving station. The images are taken as both visible light and infrared data, so you can receive satellite images even at night. (The infrared data also allows scientists to determine water and air temperature. The system is so sensitive that it can distinguish the boundaries between a weather system's cold and warm fronts.) WEFAX is the term used to refer to the broadcast of Meteorological APT/Facsimile satellite imagery.

Satellites actually transmit their images to you as a set of tones or audio impulses over certain frequencies. High tones represent light areas; low tones represent dark areas. Your radio receiver picks up these tones as the satellite orbits over your region. As the tones are received, they are transmitted by means of an earphone jack into your computer. Using a demodulator that comes with the appropriate computer software, your computer translates these audio tones into light and dark bits and produces a complete picture once the transmission is finished, usually in less than 15 minutes. You receive a satellite image of Earth in black, white, and shades of gray. You can colorize your APT image if you use a computer with the appropriate software.

The National Oceanic and Atmospheric Administration (NOAA) currently has four weather/environmental satellites in polar orbit, designated as NOAA 9, 10, 11, and 12. The former Soviet Union has nine similar satellites designated as Meteor 2-

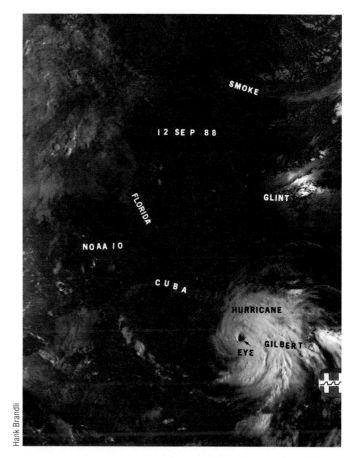

The 450-mile high NOAA 10 transmitted this near-infrared image of 1988 Hurricane Gilbert. Smoke indicated came from fires in distant Yellowstone National Park.

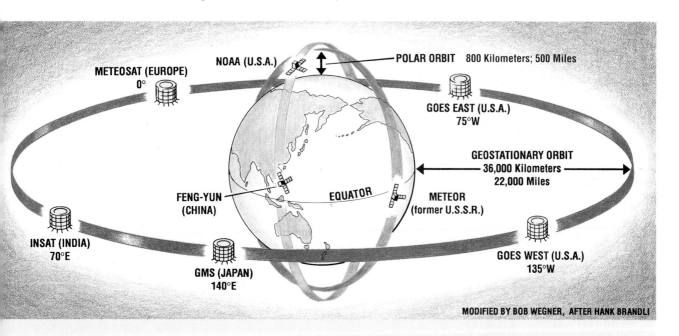

Two types of meteorological spacecraft orbit the earth with different orbital paths. From an altitude of approximately 300 kilometers (180 miles) polar-orbiting satellites move from pole to pole, scanning with both visual and infrared sensors. Geostationary satellites remain in a fixed position about 36,000 km (22,000 miles) above the equator, photographing an entire hemisphere continuously with their sensors as they follow the earth's rotation. They also relay data to earth-based receiving stations from the polar-orbiting satellites.

17, 2-18, 2-19, 2-20, 3-2, 3-3, 3-4, and 3-5. (Only a few of these satellites are operational at any given time.) The Meteor satellite is cylindrical, weighs about 2,000 kilograms (4,400 pounds), is about 2 meters in diameter (7 feet) and about 5 m high (16 feet). The Meteor has several large solar panels which provide the satellite with power. Most satellites of this class have both visible and infrared imaging systems.

These satellites and many others transmit on frequencies between 136 and 138 megahertz (MHz) in the FM band (one megahertz equals one million cycles per second, or a million hertz). Anyone with the proper equipment may acquire facsimile pictures from these satellites. The APT signal from polar-orbiting satellites — heard as audible tones of chirps, clicks, and beeps — is continuous. If you image the transmission, it produces a strip of pictures. Direct reception is limited, however, to line-of-sight. This means your ground station receives pictures from the satellite only while it is above the station's horizon. The maximum reception time of a signal, usually between 10 and 15 minutes, depends on your geographic location. During this time, a ground station receives an image that covers an area about 4,000 km long and 1,500 km wide (2,400 by 900 miles). The present frequencies of APT satellite transmissions are listed in Table 1.

American and Chinese satellites produce two side-by-side images: one in visible light, the other in infrared. The Soviet Meteor satellites produce a single image using visible light during the day and infrared at night. Meteor satellites produce a high-resolution image as a result of this single-frame imaging technique because the signal strength is not divided. The power output of these satellites is from 3 to 5 watts, which is actually about the same transmitting power of a typical CB radio. The signal seems weak, but, as the transmissions come from space directly to Earth, there are no barriers between the transmitter and the receiving antenna. So the signal, despite the low power, is heard loud and clear.

Other polar-orbiting satellites are the surveillance type. Some, such as the Soviet OKEAN series, are used for oceanic research. You will find it difficult to receive orbital information and the frequencies of American surveillance satellites, but quite easy to obtain the same information from Soviet satellites.

There is a drawback, however. The majority of transmissions from surveillance satellites are encrypted and difficult for the amateur to decode. Occasionally, these satellites do transmit using an APT/WEFAX signal and can be imaged by the simplest ground-based equipment. Surveillance satellites transmit with more power using a high-resolution imaging system, so it's worth the extra effort to receive them. Soviet satellites of this class are called the Kosmos (or Cosmos) series. Some of these satellites are now being used for scientific and other non-classified assignments rather than military missions. In Table 2 is a list of other polar-orbiting satellites received and imaged by amateurs.

Geosynchronous orbiting environmental satellites (or GOES) transmit at a frequency of 1,691 MHz. Because this frequency is in the microwave band, direct reception is not feasible. You must use a frequency downconverter to change the signal to 137 MHz before it reaches your radio. After you convert the signal, you can use the same receiver for both the GOES satellites and the polar-orbiters.

Because GOES satellites are stationary, they are very useful during hurricane season, as a storm can be watched continuously. GOES satellites are also

Table 1

Satellite	Country	Frequency
NOAA 9	USA	137.62 MHz
NOAA 10	USA	137.50 MHz
NOAA 11	USA	137.62 MHz
NOAA 12	USA	137.50 MHz
Meteor 3/2	USSR	137.850 MHz
Meteor 3/3	USSR	137.850 MHz
Meteor 3/4	USSR	137.300 MHz
Meteor 2/18	USSR	137.300 MHz
Meteor 2/16	USSR	137.850 MHz
Meteor 2/17	USSR	137.400 MHz
Meteor 2/19	USSR	137.850 MHz
Meteor 2/20	USSR	137.300 MHz
*FY-1b	China	137.795 MHz

* As of the writing of this article FY-1b had a system failure and is no longer transmitting an APT signal.

Table 2

Satellite	Country	Frequency	Remarks
Kosmos 1602	USSR	137.275 MHz	Scientific
Kosmos 1500	USSR	137.400 MHz	Military?
Kosmos 1766	USSR	137.400 MHz	Military
Kosmos 1151	USSR	51.00 MHz	Scientific
Okean 1 + 2	USSR	137.400 MHz	Scientific

useful in relaying images from polar orbiters out of range of the receiving station. For example, during the Gulf War, GOES 7 transmitted many images of the Middle East with a focus on Saudi Arabia. NOAA 11 obtained the images but, because the NOAA satellite was below the horizon of the receiving station, the image was first transmitted to GOES 7 (or another geosynchronous satellite), and then transmitted to the receiving station.

ABOVE: 1990 Hurricane Trudy, left, and a second tropical storm imaged from NOAA 11 data.
RIGHT: Because of their low altitude, polar-orbiting satellites view land and water with fine resolution. Here the cool Florida peninsula contrasts the warm waters, and the warmer Gulf Stream, flowing north at 5 to 10 miles per hour, appears black.

At present, the United States has placed five GOES satellites in orbit. The European Space Agency (ESA) has two operational GOES-type satellites called METEOSAT. All the GOES satellites listed in Table 3 belong to the United States. During morning transmissions METEOSAT relays images of Europe, the Middle East, and Africa to the United States through GOES 7.

As this article goes to press, only two GOES satellites are operating. Both send images for most of the

Table 3

Satellite	Frequency	Remarks
GOES 2	1691 Mhz	On Standby-operational
GOES 3	1691 Mhz	Off-Standby
GOES 5	1691 Mhz	Off-Out of fuel
GOES 6	1691 Mhz	Operational
GOES 7	1691 Mhz	Operational
GOES 1	1691 Mhz	Launch date Mid-1992/replacement for GOES 5
TIROS-N		Images relayed through GOES 7

day. If you live in the western part of North America, you tune to GOES 6 (also known as GOES West). If you are in the central part of North America, then you use GOES 7. On the East Coast you also tune to GOES 7, because GOES 5 (East) is now turned off. (In 1992, stations on the East Coast will be able to receive GOES 1.)

Putting Together Your Earth Receiving Station

To assemble your earth-based satellite receiving station you need the following equipment: a receiver, antenna, low-loss antenna cable, and — if you plan to image (or digitally capture these images) — IBM computer (or an IBM clone), and the appropriate software.

The Receiver. The important environmental satellite frequencies are located in the 2-meter band. The VHF-UHF police scanner radio operates in this range and can be programmed for as many as 100 frequencies. Although the bandwidth of a stock receiver is not suitable, it can be modified by a simple adjustment in the receiver's IF circuit to obtain a bandwidth between 30 and 50 kilohertz (KHz). (A kilohertz is a thousand cycles per second.)

The adjustment costs only a few dollars, and it is best to have this procedure done by a qualified radio technician. Many scanners can be purchased wholesale from radio dealers for about $50. If you want to buy one already modified, you will probably pay from $100 to $175. The advantage of the programmable scanner is that you can add or delete frequencies at the push of a button. I use a 16-channel Bear Cat scanner/receiver with the bandwidth set at about 45 KHz. This receiver provides excellent coverage of polar-orbiting satellite transmissions. Some scanner receivers need no adjustment. I obtained excellent results from receivers with a bandwidth from 20 to 50 KHz. Check your receiver specifications in its instruction book.

Vanguard Labs, of Hollis, New York, has designed a receiver exclusively for satellite reception and imaging. Their model WEPIX 2000 scans over eight popular satellite frequencies. This receiver is very good and provides excellent reception of satellite transmissions, but it does fall short in versatility. The frequencies are locked into the receiver and can't be changed. Unlike the scanner radios, the Vanguard WEPIX 2000 limits you to frequencies provided by the manufacturer (present cost of the Vanguard WEPIX is about $330).

Almost any radio that receives in the frequency range listed above will work well. You can spend as little as $50, or you can purchase a professional receiver such as the Kenwood Model RZ-1 for $500. Or you can be really extravagant and get the ICOM Communications Model R-9000 for just under $5,000. But remember, to achieve good results as you receive and image these satellites, you don't need to spend more than $75 for the receiver.

Receiver Preamplifier. A receiver preamplifier can be used to improve the signal-to-noise ratio of very weak satellite signals. The preamp is normally placed between the antenna and the receiver. I found the most efficient location to be up near the antenna. Normally the preamp requires a 12-volt power supply sent by the receiver through the antenna cable. Some receivers, such as the Vanguard WEPIX, have connections for the preamp. Others, such as the ICOM Communications and Kenwood models mentioned above, already have a built-in preamp. A preamp also benefits the VHF-UHF scanners, but unless you use more than 23 m of antenna cable (75 feet), you won't need it because these receivers are very sensitive.

The Antenna. Selecting the proper antenna is very important. Because most satellite reception takes place near the 2-meter amateur band, you have a wide selection from which to choose. You can take advantage of this availability or build your own antenna and save as much as $100. First, let's discuss four antennas used for receiving and imaging the polar-orbiting satellites. Then we'll cover the reception and imaging of the GOES-class satellite using a loop yagi antenna. Unless otherwise specified, try to mount your antenna in as high a location as possible, usually the roof.

The yagi, or beam, antenna is a typical TV antenna with multiple elements. Many models are available for the 2-meter band, ranging in price from $100 to $500. The more elements on the boom of the mast, the better the reception and the more directional the antenna will be. This antenna provides maximum signal gain but is directional, which means that the antenna will have to be moved at least twice to keep it pointing at the satellite.

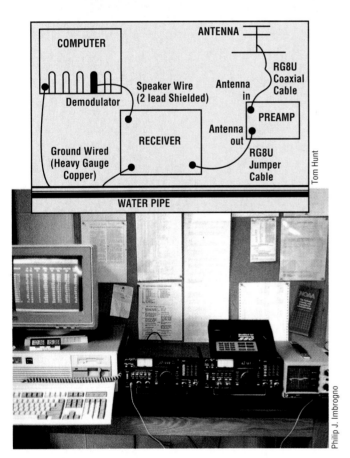

Diagram, above, shows simple schematic of a satellite receiving station, below, assembled from readily available components.

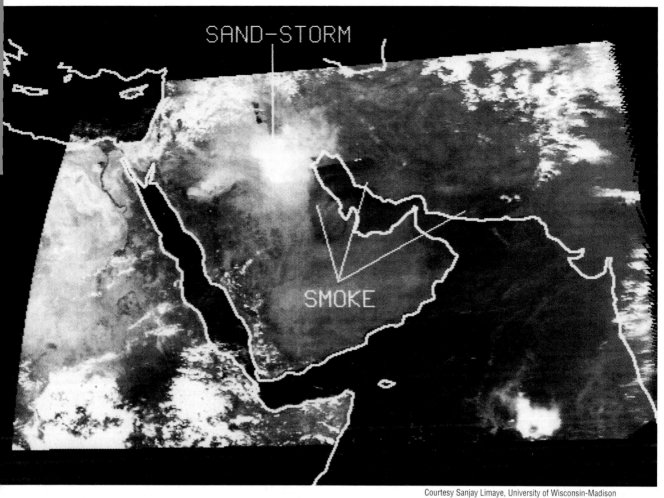

A sandstorm and oil-fire smoke cloud the atmosphere in this May 1991 image of Saudi Arabia, relayed from Meteosat to GOES 7.

These antennas work quite well and moving the antenna is a minor inconvenience when you consider the signal's strength compared to that of other antennas. You don't need to buy an expensive amateur radio yagi; you can use a modified TV antenna. I obtained good results with a five-element TV antenna cut down to receive 2-meter band signals. The antenna elements should be in multiples of two meters (about 6 feet): 0.6 m for one-third-wave (2 feet), 1 m for half-wave (3 feet) and just under 2 m for full-wave (6 feet). The closer you get to the wavelength of two meters, the better the reception will be. I cut the TV yagi beam elements to one-third-wave for 137.5 MHz (exactly 0.66 m or 2.18 feet) and obtained excellent results.

The dipole is a simple strand of heavy-gauge copper wire stretched horizontally. It might be the antenna for you if you are on a tight budget. The wire is divided in half by a plastic insulator so you can attach both ends to the antenna signal wire and shield wire. Because polar satellites orbit north to south and south to north, the antenna should be oriented east to west and mounted about 50 cm (21 inches) off the ground. The dipole antenna provides good overhead coverage but won't pick up satellite transmissions low on the horizon.

The discone antenna is designed to receive over a very broad range and is often sold for VHF-UHF scanner/receivers. This antenna performs well when the satellite is low on the horizon. As the satellite passes overhead, the signal rapidly deteriorates. To solve this problem, place a wire mesh about 60 cm (two feet) under the antenna. This reflects the signal up to the antenna as the satellite passes overhead. The discone antenna takes little room and is very economical.

The Turnstile Reflector WEFAX Antenna is a custom-made antenna offered by Vanguard Labs. This one-third-wave antenna gives you 120° of omni-directional coverage. The four active elements of the antenna are tuned to cover the 136-138 MHz band, and eight lower antenna radial elements provide a very effective ground plane. Reflector elements at the base of the antenna insure that the satellite will not fade out as it passes overhead. This antenna is light, compact, and available with or without the preamp.

Antenna Cable. Low-loss cable such as RG 58U or RG 59U should be used in lengths of no more than 23m (75 feet). The connectors are standard PL-259s on both ends which mate to the bottom of the antenna and then to the back of the receiver. (PL-259 is the typical connector used for both CB and amateur radio.)

Receiving GOES Satellites

Geosynchronous satellites give us a very large view of the entire earth, producing the same image you see during the weather report on your local TV newscast. Occasionally, polar-orbiting satellite images from all over the world are relayed through a GOES satellite belonging to the United States or Europe. At present, GOES 7 and sometimes GOES 6 are the only operational satellites of this type in the western hemisphere.

The GOES-type satellites transmit on a frequency of 1,691 MHz, which is a microwave signal. The antenna receiving this signal is usually a dish with a minimum diameter of 60 cm (2 feet). Dish antennas can be quite expensive and difficult to maintain. I suggest that for starters you try an antenna called the loop yagi, just 10 cm in diameter (4 inches) and almost 3 m long (about 9 feet). It weighs only a little over a kilogram (about 2 pounds) and can be mounted on an adjustable camera tripod head. The tripod head should be adjustable, as you will have to move the antenna occasionally to keep it pointing toward the satellite. If you need more signal gain, an extension can be added to the loop yagi to increase its signal gain by over 50 percent.

You feed the loop yagi antenna into a 1,691 MHz downconverter using no more than 1.5 m (5 feet) of low-loss antenna feedline. (NOTE: The connection between antenna and input of the downconverter must be made with low-loss Belden 9913 coaxial antenna cable because regular cable will not transmit microwave signals. The cable connection between the output of the converter and the receiver can be made with standard RG 59U cable.) The downconverter changes the microwave signal to 137.5 MHz so it can be picked up by your receiver if set at this frequency. The downconverter runs off a 12-volt power supply. If the unit is kept outside, it must be placed in a weather-proofed box.

The GOES satellites orbit above the equator, which places them south of all stations in the Northern Hemisphere. To roughly calculate the elevation of a GOES satellite above your horizon, subtract your latitude from 90°. For example, if your latitude is 40° north, then the satellite will be about 50° above your horizon. Using slow sweeping motions, point your antenna to the southwest if you are east of the satellite; point your antenna to the southeast if you are west of the satellite. You should have no trouble finding the GOES signal which produces audible clicks and chirps on your receiver.

Equipment for receiving GOES satellites is the most expensive part of this project. You use the same receiver that you use for polar-orbiting satellites, but the antenna and the downconverter cost approximately $600. Table 4 contains the present longitude of the two operational GOES satellites. Choose the one nearest you.

SUPPLIERS

SOFTWARE SYSTEMS CONSULTING, 150 AVENIDA CABRILLO, SAN CLEMENTE, CA 92672, Telephone: (714) 498-5784. Imaging and orbital predication software, scanning satellite receiver, loop yagi microwave antenna. PC HF Facsimile software and demodulators.

AMATEUR ELECTRONIC SUPPLY, 5710 W. GOOD HOPE ROAD, MILWAUKEE, WI 53223, Telephone: (414) 358-0333. Shortwave receivers, VHF-UHF receivers, antenna systems, antenna cable.

QUORUM COMMUNICATIONS INC., P.O. BOX 277, GRAPEVINE, TX 76051, Telephone: (817) 488-4861. Satellite downconverters, GOES antenna systems, antenna cable.

VANGUARD ELECTRONIC LABS, 196-23 JAMAICA AVE., HOLLIS, NY 11423, Telephone: (718) 468-2720. Custom satellite receivers, satellite antenna system for polar-orbiting satellite, preamplifiers, imaging and orbital software.

RADIO SHACK (local throughout United States): Antenna systems for 2-meter and 10- to 20-meter antenna cable.

The phone numbers below are computer bulletin boards for information on satellite news, reception, imaging, and orbital predictions. To access the information, you need a computer, communications software, and a modem. Use of the bulletin boards is free of charge.

NOAA/NESDIS	(800) 546-1000
Vanguard Labs	(718) 740-3911
Datalink	(214) 394-7438
Celestial BBS	(513) 427-0674
NASA Spacelink	(205) 895-0028
Software Consulting	(619)-259-5554

Nation-wide computer services like COMPUSERVE also supply information on satellite reception and observation. These services do charge to access their data banks. COMPUSERVE operates a toll-free number for information: (800) 848-8990.

Table 4

GOES 6	135 degrees West
GOES 7	93 degrees West
GOES 1	65 degrees West

Imaging Environmental Satellites

You image satellite transmissions using an IBM PC or PC clone computer. Your computer should be an MS DOS 2.1 version or higher with at least 512K bytes of memory and must have either an EGA, VGA, or SVGA graphics card (or one that is compatible) to handle the complexities of forming the graphics image. The signal feeds into the computer from your receiver through an earphone jack (or demodulator) plugged into a serial port at the back of the computer. (See diagram on page 62) You also need a color monitor that supports VGA images. You can use a black and white monitor but you won't be able to colorize your images and they won't have the higher resolution provided by a color monitor.

Several programs on the market will track and image satellites. One of the easiest to use is PC GOES/WEFAX by Software Systems Consulting. With this program, you can track satellites and predict when they will pass within range of your station. The software comes with the name, frequency, and orbital elements of eight polar-orbiting satellites stored on the disk. Every two months or so, you need to update the orbital elements, an easy task because all you do is enter the new values into the program. You can obtain these updated orbital values from a number of computer bulletin boards that provide satellite information. (See the box on page 64 for telephone numbers.)

The demodulator (supplied with the software) plugs into the speaker, or earphone, of the receiver and then into the back of the computer serial port. It decodes the satellite tones and produces the image on the computer screen. The PC GOES/WEFAX software with the demodulator costs about $250.

The National Weather Service and the U.S. Navy transmit satellite images of the Western Hemisphere on the shortwave radio band seven times in a 24-hour period. These high-quality and high-resolution transmissions are re-broadcast from GOES images. Recently, while listening on the shortwave band, I received high-resolution satellite images from an unknown Soviet ground station. These images probably originated from one of the polar-orbiting satellites, then were re-broadcast from the Soviet ground base. In the shortwave band between 3 and 30 MHz are thousands of frequencies; hundreds are used by the governments of the world to transmit information, maps, pictures, and satellite data. To image these transmissions, you need a shortwave receiver that tunes into the upper and lower sideband, a shortwave antenna (a long strand of copper wire will do) and an IBM PC computer (as described above). Your PC GOES/WEFAX software and demodulator will also image these shortwave Fax transmissions.

In Table 5 are listed some popular frequencies and the location of the transmitting stations. In addition to satellite images, they also transmit

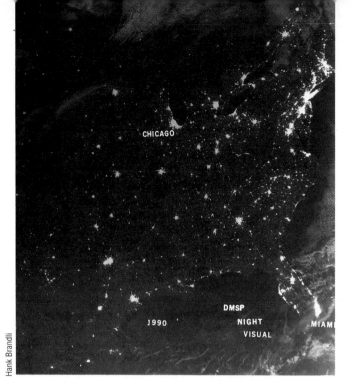

October 1990 image from Air Force Defense Meteorological Satellite Program. Night lights clearly indicate cities of the eastern United States, Canada, and Mexico.

weather charts. You may be able to obtain a free list of transmitting stations and their schedules if you write to The Superintendent of Documents, U.S. Government Printing Office, Washington, D.C. 20402 and ask for Worldwide Radio Broadcast Schedules.

Table 5

Location of Station	Frequencies
Norfolk, Virginia	8080, 10865 KHz
San Francisco, California	8682, 12730 KHz
United Kingdom	2618 KHz
Minsk, Belarus	3810 KHz
Halifax, Nova Scotia, Canada	4271 KHz

Endless possibilities exist for imaging these satellites. The earth station described in this article is now set up at the Windward School in White Plains, New York. The station, used to teach students about our planet by studying it from space, generates much excitement in both students and their parents. Because of its versatility and the global perspective it provides, satellite imaging has become a powerful teaching tool. ⊕

Philip J. Imbrogno is the science coordinator at the Windward School in White Plains, New York. He put together a complete earth station that receives and images more than 25 environmental satellites.

THE DEADLY SWEPT POWER OF AWAY TSUNAMIS

Courtesy U.S. Navy

The aftermath of a tsunami that struck Hilo, Hawaii, in 1960 testifies to the destructive force of the waves. This tsunami, generated by an earthquake off the coast of central Chile, affected the entire Pacific Basin.

by John Dvorak and Tom Peek

When an earthquake struck 30 miles off the Pacific coast of Nicaragua last Sept. 1, many people living in the tiny seaside villages didn't feel the shaking. But they did notice that low tide seemed to have come at a peculiar hour. The sea had retracted, as if it were holding its breath.

Minutes later, the sea exhaled. Waves as tall as four-story buildings thundered toward beachfront houses and hotels. As each wave approached, the ocean swelled into a long ridge and then surged toward 200 miles of coastline, crushing houses and sweeping people out to sea. The waves destroyed or damaged thousands of homes, displacing about 14,500 people. More than 150 people were killed or lost, an overwhelming majority of them children who were washed away while they slept.

Today, hundreds of millions of people live in coastal communities vulnerable to tsunamis ("harbor waves" in Japanese). To unsuspecting coastal dwellers, an onslaught of these giant waves comes as a freakish tragedy — a rare perturbation in the short span of human life. But in the context of geologic time, tsunamis are expected natural events, an inevitable consequence of the restless movement of Earth's tectonic plates.

The likelihood of devastating tsunamis varies widely, but the majority occur in and around the Pacific rim, where dozens of major earthquakes and volcanic eruptions occur each year. When one of these events disturbs the ocean floor, sea waves may race across the Pacific, some of them bringing disaster thousands of miles away.

Since 1946, the year a tsunami destroyed one-third of Hilo, Hawaii, geophysicists have used warning systems to alert the public to the approach of giant waves. But tsunamis generated close to shore can assault the coast almost without warning. And ever-increasing coastal development complicates evacuation plans, demanding a vigilant and educated public. In light of this, are we ready for the next big wave?

Survivors of the Nicaraguan tsunami of Sept. 1, 1992, sit among wave-wracked debris. It was triggered by a magnitude 7.0 earthquake off the Pacific coast and killed more than 150 people.

Geophysicists only partially understand how large tsunamis form. One reason is that these events are tricky to study: They occur infrequently, and few people have ever directly witnessed or measured their formation. But researchers agree on the general causes of tsunamis, which include landslides, undersea earthquakes and volcanic eruptions. Occasionally, gigantic waves have even been unleashed by asteroids that plunged into the sea.

Historically, undersea earthquakes have triggered most tsunamis, but volcanic eruptions and landslides have produced the highest waves. Such quakes are most likely to occur along the subduction zones of the Pacific rim, where the ocean floor is being shoved beneath the edge of continents and island chains. Alaska, Japan and the western coasts of Central and South America are perched on such subduction zones, which is why some of the world's most powerful tsunamis occur there.

Only certain types of undersea earthquakes typically cause tsunamis. For example, the horizontal shifts that occur along strike-slip faults like the San Andreas, which runs offshore in places along the California coast, leave the ocean floor relatively undisturbed. But if the seafloor is abruptly dropped or raised, water is displaced and the possibility of a tsunami is high. This is what happened during a major earthquake on March 27, 1964, in Anchorage, Alaska. The quake hoisted 35,000 square miles of seafloor and islands (equal in area to the state of Indiana) as much as 50 feet. The sudden upward movement also raised the ocean layer above the seafloor, and the water poured away in a series of long waves.

How undersea volcanic eruptions produce tsunamis is still debated. The 1883 eruption of the volcanic island of Krakatau, located in the Sunda Straits between Sumatra and Java, Indonesia, generated huge tsunamis whose waves reached as high as 150 feet. Though Krakatau was uninhabited, the waves drowned about 36,000 people on nearby islands; more people have died as a result of these tsunamis than any others in recorded history.

Until the 1960s, most scientists thought the deadly tsunamis were produced by a sudden drop of the seafloor, triggered when the roof of a subterranean magma chamber collapsed as the magma erupted. But scientists rejected this notion once they considered the initial movement of the waves. A sudden drop that disrupted the ocean would have first pulled water toward the volcano, and someone on one of the nearby islands would

have seen the sea retreat. Instead, survivors reported that the arrival of the waves was initially marked by a rise in sea level. The record from a sea-level gauge 200 miles from Krakatau confirmed their observations. Now scientists believe that debris from the volcano displaced the water, either by flowing along the seafloor or by plummeting to the sea after it was ejected.

As awesome as the tsunamis were at Krakatau, waves ten times higher, by far the largest on record, can be produced when landslides fall into narrow water inlets, such as the fjords of southern Alaska and Scandinavia. In 1958 one such wave overwhelmed three fishing boats in Lituya Bay, along Alaska's Inside Passage. A nearby earthquake triggered a rock and ice avalanche from the steep walls of the bay, causing water to surge up the opposite side to a height of 1,740 feet, more than three times the height of the Washington Monument. The crews of two fishing boats survived; the third crew, which attempted to outrun the wave, was never found. (Technically, the fatal surge in Lituya Bay and those at similar inlets elsewhere are not tsunamis but "swashes," because they more closely resemble splashing water than the movement of individual waves.)

Yet even these surges seem insignificant when compared with the stupendous walls of water that must be produced when an asteroid, traveling several miles per second, plummets into the ocean. Simply by happenstance, every million years or so a mile-wide asteroid hits Earth. Because three-quarters of our planet is covered by seawater, asteroids are most likely to land in the water, creating enormous tsunamis that drown islands

In the deep ocean, tsunamis scarcely raise the sea surface. But unlike the crests of ordinary waves, their crests can be 10 to 100 miles apart, much longer than the ocean depth. They grow tall when they encounter resistance from the seafloor near shore. Friction slows the front of the wave, while the back swells into a wall of water.

Water is sucked out to sea before a tsunami inundates the land surrounding Oga Aquarium in Akita Prefecture, Japan, on May 26, 1983. Water first pulls away from shore (far left). The waves eventually reach their maximum level, setting a car afloat (near left) and then completely withdraw (below). This tsunami drowned more than 100 people in Japan and three people in Korea.

and scour continental shorelines. Geologists have realized this for many years but have only recently uncovered evidence of such catastrophes.

In southeastern Texas, a thick deposit of mudstone laid down millions of years ago on the bottom of a quiet, shallow sea is interrupted by a two-foot-thick layer of sandstone containing meteorite fragments. These fragments lie at the 65-million-year-old Cretaceous-Tertiary boundary, the time of the great worldwide extinctions and the end of the dinosaurs. Many scientists now believe that the after-effects of an asteroid impact caused the extinctions. A 100-mile-wide crater buried beneath the Yucatan Peninsula is thought to be the site of the impact. Apparently the asteroid struck an ancient Caribbean Sea and produced huge tsunamis, high enough to expose the deep seafloor and inundate the continental coastline. The backwash pulled plants and rocks into the sea where they are now found in the sandstone layer in Texas and in similar meteorite-rich layers in Mexico and Haiti.

Similar evidence suggests that another large asteroid impact 40 million years ago may have produced sea waves hundreds to thousands of feet high that scoured the coast of the land that is now Virginia and Maryland. The crater has not been found, but it must lie in the Atlantic Ocean near the coastline. Geologists will likely uncover evidence of more such cataclysmic events, gaining a better understanding of the effects of these collisions on the oceans and distant coastlines.

Far out in the ocean, tsunamis raise the sea surface only a trifling few inches. As conservationist Wallace Kaufman and geologist Orrin Pilkey write in their book, *The Beaches are Moving*, "Travelers in the open stretches of the Pacific Ocean have ridden out the most powerful waves ever known without putting down their cocktail glasses." But the same tsunamis that pass subtly, even imperceptibly, beneath ships at sea can be transformed into towering walls of water as they near a shore.

Tsunamis are fundamentally different from the waves that normally break along the shore. Ordinary waves, caused by wind and distant storms, take several weeks to cross an ocean and have wave crests spaced a few hundred feet apart or less. Tsunamis, by contrast, are called long water waves. Their crests can be tens, even hundreds of miles apart, much longer than the ocean is deep. As a result, tsunamis travel hundreds of miles per hour (the speed of commercial jet airliners) and cross the Pacific Ocean in a single day.

A tsunami metamorphoses into a wall of water when the front of the wave encounters friction on the shallow seafloor near shore. The resistance slows the front of the wave, but the back of the wave, still in deep water, continues to move quickly. The long, deep wave swells into a tall ridge. Instead of breaking on the beach, the tsunami plows ashore, flooding vast areas and potentially causing immense destruction.

This dynamic was tragically illustrated on June 15, 1896, when a 90-foot-high tsunami hit Japan, causing about 27,000 deaths. Earthquakes are common in Japan, so the people of the seaport Kamaishi were not alarmed when the ground began to sway slowly,

THE WAVES OF '46:
A SURVIVOR REMEMBERS

Kapua Heuer recalls the tsunami of 1946 from the cliff above Hilo Bay where she and her daughter watched the waves that devastated parts of the city.

"Mamma, why is there no water in Hilo Bay?" asks the little girl peering out the window of her house atop a cliff. The girl tries to ready herself for school but is distracted by the uncanny view out her window. The arm of the sea that fills three-mile-wide Hilo Bay has pulled back, stranding hundreds of fish that flop madly on the drained seafloor. The girl watches amazed as underground streams, usually held back by the pressure of seawater, spout from the exposed floor of the bay. Kapua Heuer joins her daughter at the window. They and the rest of the family have no idea that the cause of the spectacle is a tsunami about to strike.

Kapua and her daughter race outside and stand on the edge of a 30-foot precipice. The tsunami, born six hours earlier in Alaska, was sweeping at nearly supersonic speed across the Pacific. As the steel gray waves roar into Hilo Bay, they swell to enormous heights, killing 96 people and leaving the coastal stretch of the city in ruins. Fortunately, Kapua and her family are safe atop the cliff as a dozen or more gigantic waves begin to rise and fall across Hilo. "It looked like a whole mountain coming in," 80-year-old Kapua now remembers . . .

With a resounding crash, the first wave collides with outgoing water sucked from the shore. The waves lap over the top of the cliff, roll into the city and plow a row of bay-front buildings into those across the street. From Kapua's perch a half-mile away, the impact sounds oddly like a giant bag of potato chips crushed all at once.

A strange sucking sound fills the air. Again, the bay begins to drain, this time carrying with it seven or eight people. Their heads bob about as they frantically try to cling to chunks of wood and other debris seized by the wave. Kapua recalls: "As the suction is going out, there is another wave forming and coming in, and again the two meet with a terrible crash. When the second suction comes out, there's no human life struggling out there. They're gone!"

Kapua and her family move from the cliff as the second wave hits the city. Then a third time the bay empties. A freighter runs aground inside the breakwater built to protect the bay. When the next wave passes, the ship steams out and escapes into the deeper water beyond the disintegrating breakwater.

Kapua drives her children to Hilo. They inch past demolished homes, stunned at the sight of an elderly neighbor washed up naked near the road. Kapua drops the girls at their school and continues to the U.S. Navy facility where she works. The lieutenant on duty asks her and Phil Brown, another employee, to drive to a settlement along the coast to check on conditions at a hospital.

As they thread their way through the devastation, Kapua and Phil realize that a third of Hilo has disappeared. Battered automobiles look as if they'd been beaten in a gigantic mixer. Railroad cars are overturned, the train station is gone and the trestle bridge across the Wailuku River has been swept several hundred feet upstream. Small boats and sampans lie upturned on the beach. Coastal communities are wiped out, people and homes pulled into the sea.

Deep water across the road forces Kapua and Phil to abandon their car. They borrow a horse from a resident, but eventually the water forces them off the animal. To reach the hospital, they swim through a flooded woods where they see an arm reaching out of the seawater, the last signal of a dead man wedged between trees. Shocked and exhausted, Kapua and Phil drag themselves onto dry land and steady their nerves with a bottle of wine they find floating in the debris. Fortunately, they learn some good news on this grim morning: While the hospital has been irreparably damaged, its evacuation is under control.

The next day Kapua faces the gruesome task of identifying bodies recovered from the the sea or the wreckage on land. Bodies of a dozen people covered with blankets line the sidewalk. Kapua relates: "You lifted the blanket to see if you could find who you were looking for. The stark terror in their eyes — terror! They died in terror!" Among the victims is a friend who died when a hotel collapsed.

Even now, almost 50 years later, Kapua Heuer is struck by the destructive power of the sea. "It was unbelievable. Even in a dream, I don't think you could dream up the devastation we saw," she says. "You're looking at something today and you see it all, and tomorrow it's nothing but shambles."

— *Tom Peek and John Dvorak*

This coastal community in Oahu, Hawaii, was flooded by a tsunami on March 9, 1957. The waves were triggered by an earthquake in Alaska's Aleutian Islands.

the result of seismic waves passing from a distant earthquake under the sea. Twenty minutes after the first swaying, the sea began to recede. Forty-five minutes later, residents heard a sound like a powerful rainstorm, and a 90-foot-high wall of water poured in on the town. The Kamaishi fishing fleet was far out to sea that evening and knew nothing of the earthquake or the tsunami that had passed beneath them. When the crews returned home the next morning, they were stunned to sail through miles of wreckage and floating bodies.

Tsunamis can be deceptive in other ways as well. Their path, for instance, is often altered by the relief of the undersea landscape. Submerged mountains and valleys can disperse tsunamis in midocean. That's why giant waves that originate in Japan usually have no impact on the Hawaiian Islands: A 2,000-mile-long chain of small islands and seamounts to the northeast serve as a natural barrier. But no such breakwater protects Hawaii from tsunamis produced in Alaska or South America.

The highs and lows of the seafloor can also focus waves onto segments of the shoreline, as was tragically illustrated when the Alaskan tsunami of 1964 swept down the west coast of the United States. Four hours after the earthquake, a five-foot wave washed onto the northeast corner of Washington. Thirty minutes later, an unimpressive two-foot high wave rolled into Astoria, Oregon. But farther down the coast the sea wave, focused as it passed over an undersea volcano, rose into a 21-foot-high wave that hit Crescent City, Calif., causing eleven deaths and levelling much of the waterfront.

If determining the path of a tsunami is difficult, predicting the size of a wave in a series is nearly impossible. Tsunamis usually consist of multiple waves, and the largest may occur anywhere within the series. Many of the 61 people killed by the 1960 tsunami in Hilo, Hawaii, were unconcerned after two relatively small waves hit the city, only to be swept away or crushed by the third, a 35-foot-high wall of water.

When we envision a tsunami's effects on human lives, we tend to focus on the damage caused by the sheer force of the waves: drowned or injured people, razed buildings, undermined foundations, collapsed bridges, hurled cars. But it is not only wave action that makes tsunamis dangerous. Widespread flooding inundates power plants and oil storage facilities and swamps sewage treatment centers. The subsequent pollution can be as disastrous as wave damage. The 1964 Alaskan earthquake and tsunami ripped open waterfront oil storage tanks. The oil ignited and was carried inland on fiery sea waves as high as 30 feet. After the Nicaraguan tsunami last September, sewage contaminated wells, causing an outbreak of cholera. Traffic jams and power outages

also invariably take their toll, slowing a community's ability to evacuate and provide medical relief.

Today, several local tsunami warning systems are in place, and regional systems exist in Alaska, Hawaii, Japan, French Polynesia and Russia. Most devastating waves are generated relatively close to shore, but there is one warning system designed to monitor the entire Pacific in case of a rare, ocean-wide event.

The Pacific Tsunami Warning Center, a complex of unadorned cement block buildings, sits on a broad floodplain west of Pearl Harbor in Hawaii. In the largest building, needles scratch away on seismic drums, recording signals from 20 instruments scattered throughout the Pacific. The instruments detect the passage of seismic waves through the Earth, the telltale signature of an earthquake. Teletypes patter away in support of the drums, relaying additional information about the largest quakes.

Above these machines, a large map depicts the planet as if it were drained of seawater. Large features of the ocean floor are as clear as the more familiar patterns of mountains and plains across the continents. The map is dotted with 80 stick pins, each representing a station that records the local rise of the Pacific Ocean. These stations transmit their readings to a geostationary satellite, indicated by a pin that stands taller than the rest. If a station reports a significant earthquake or sea-level change, the center alerts emergency authorities. This work is done in cooperation with agencies in more than 20 other countries and territories throughout the Pacific region.

An alarm (a loud buzzing followed by a shrieking tone) sounds when an earthquake of a magnitude greater than 6.0 occurs somewhere in the Pacific or when an earthquake half that size occurs in Hawaii.

On March 27, 1964, Alaska experienced a magnitude 8.4 earthquake, one of the largest recorded in America and the most devastating in Alaska's history. At Seward, the quake sent a section of the waterfront sliding into the bay. Waves spread in all directions. Trucks were demolished and ships run aground. Petroleum tanks ruptured and the spilled oil ignited, setting on fire several buildings and an electrical generation plant.

Within minutes, scientists determine the location and magnitude of the quake and decide if unusually large waves have struck any sea-level monitoring stations close to the quake.

Poised in front of their computers, the scientists calculate and wait. If the stations indicate a tsunami or if the earthquake is large enough to warrant immediate concern, someone calls the National Warning Center in Colorado. Nearby is another phone (or its radio backup) used, if needed, to announce the possibility of a tsunami to Hawaii Civil Defense and emergency offices on each Hawaiian island. This is a tsunami watch.

As more data come in indicating the progress and height of the waves, the scientists decide whether to turn the watch into a warning. According to Mike Blackford, director of the center, they issue a warning when they have evidence of a tsunami at least three feet high located 500 to 1,000 miles from the quake's source. This means the tsunami is moving across the Pacific.

Blackford, a ten-year veteran of Alaska and Hawaii warning systems, says feelings are tense when these decisions are made. "We don't want to cry wolf," he says. "We don't panic. We just follow certain procedures and attempt to be objective." But the pressure is on. If a warning is issued to local

The surge wave that drove this plank through a truck tire in Whittier, Alaska, demonstrates the often bizarre nature of tsunami destruction. The 1964 tsunami killed 13 people in Whittier, a community of 70 people.

tsunami that roared into Hilo, Hawaii, in 1946 claimed 159 lives. A sailor oard the grounded ship Brigham Victory took this photograph just before e wave slammed into the man at lower left in the picture and killed him.

emergency authorities, community sirens sound the need to evacuate, radio stations broadcast emergency information and police officers and fire fighters mobilize the public.

The warning center is useful for oceanwide disturbances, but smaller tsunamis generated closer to shore can hit with almost no warning. These waves often catch the public off-guard. "It takes about a half-hour to get information from warning centers to the state authorities, and sometimes that's all the time you have before the tsunami hits," says Jane Preuss, a land-use planner at Urban Regional Research, a Seattle consulting firm.

Even when authorities have enough time to prepare, evacuation is a challenge. Coastal development continues at a rapid rate, so planners must use novel means to remove people to safe areas. In Honolulu, for example, where tens of thousands of residents and tourists work and play along sandy beach-fronts each day, emergency plans call for people at street level to evacuate upward into high-rise hotels and condominiums. Some buildings in Hawaii and Japan were designed with tsunamis in mind. The lobbies of some major hotels are nearly wide open so a wave will not expend its force on the building, but will pass through unimpeded. Even so, in areas unaccustomed to dealing with tsunamis, like coastal Nicaragua, there may be few, if any, plans for dealing with such a disaster. And in tsunami-prone areas that have not been hit recently, developers and residents may unwisely deny the risks.

One of the biggest problems is the public percep-

tion of "false alarms," warnings issued for tsunamis that become insignificant by the time they reach a coastline. This undermines confidence in the warning system. After several false alarms in the 1950s, some Hilo residents ignored the 1960 evacuation order and were killed watching for the waves. In 1964, 10,000 curious people crowded the coastline near San Francisco, waiting for the predicted tsunami, which materialized 200 miles north and pummeled Crescent City. In addition to their numbing effects on the public, false alarms are expensive. A 1986 evacuation in Honolulu, the result of a warning following a 7.7-magnitude earthquake in Alaska, cost $30 million. When the waves rolled in, weakened by winds and storms, they were only inches high.

But Blackford insists it is better to err on the side of caution. By incorporating better models of sea-floor movement into warning systems, scientists will reduce the possibility of false alarms, he argues. Such models would help them predict wave heights more precisely, something that has thus far been what he calls "a Monday morning quarterback exercise."

Of course, public awareness is most important of all. Blackford says people just have to know what to do. "If you're in a coastal area and you're involved in an earthquake and you cannot stand up, as soon as you *can* stand up, head for the highest ground you can find. If there is no high ground, run — don't walk — away from the water." ⊕

John Dvorak is a research geophysicist and avid scuba diver. Tom Peek is a novelist living in Hawaii who has enjoyed untold hours watching lava flow into the sea from Kilauea Volcano.

CHARTING EARTH'S FINAL FRONTIER

By Tom Yulsman

EAST PACIFIC RISE

The major features of the ocean floor, as seen in the large map, have been well known for decades. Now sonar is revealing the details, such as the topography of the East Pacific Rise (above), part of the midocean ridge system.

NASA's Magellan spacecraft has imaged and mapped 99 percent of the surface of Venus, revealing surface textures and topographic features that rise as little as 160 feet above the surrounding landscape. In contrast, only 10 percent of Earth's own deep ocean floor has been surveyed in the same detail.

But that percentage is slowly growing, thanks to major advances in sonar technology and intensifying survey efforts. And as scientists scrutinize more and more of the seafloor, they're making some amazing discoveries.

In February, researchers from Brown University and the University of California, Santa Barbara,

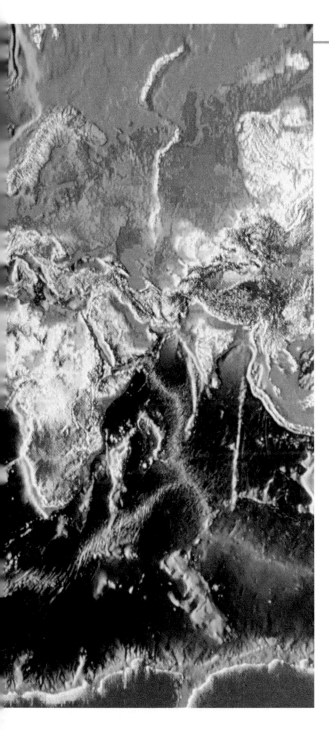

announced that they had found 1,133 volcanoes clustered in an area of the South Pacific about the size of Washington state — the greatest known concentration of volcanoes on Earth. The scientists made the discovery while surveying the seafloor with two sonar devices: a sidescan sonar, which produces images that reveal textural details like lava flows, and a multibeam swath-mapping system, which reveals topographic features like volcanic cones.

The volcanoes lie 2,000 miles west of Peru near the East Pacific Rise, part of the globe-encircling midocean ridge system. Magma from the mantle creates new crust as the seafloor spreads apart at the ridge system. According to Ken Macdonald of U.C. Santa Barbara, the volcanoes may be linked to massive currents of hot rock believed to be rising within the mantle under the East Pacific Rise.

How did more than 1,000 volcanoes between 2,000 and 7,000 feet high remain hidden until now? Until the 1970s, sonar could not determine the shape of features on the ocean floor. It was only capable of determining the depth of the water over the highest feature within a patch of seafloor. Thus, scientists had difficulty telling the difference between a volcano and a ridge. With the advent of modern sonar systems, these technological limitations vanished. Even now, however, detailed mapping is hampered by the fact that only four university-operated research ships are properly equipped to do the job.

But scientists have made progress. Using a sidescan sonar device called GLORIA (Geológic Long-Range Inclined Asdic), the U.S. Geological Survey imaged all the ocean floor in the U.S. Exclusive Economic Zone, an area that extends 200 nautical miles off the coasts of the United States. GLORIA emits sound waves to either side of the towing ship's track, covering a swath as wide as 36 miles. Based on the intensity of sound returning from the seafloor, a computer produces images that reveal structural and textural details. By towing the device in a pattern akin to mowing a lawn, scientists can image a region the size of New Jersey in just 24 hours. This makes GLORIA ideal for reconnaissance of features warranting a closer look with other technologies, says USGS geologist Jim Gardner.

Images from sidescan sonars help scientists identify faults, channels, canyons, volcanoes, lava flows, landslides and debris fields. Among many findings in the exclusive economic zone, USGS scientists identified fields of rocky debris covering thousands of square miles off some of the Hawaiian Islands. This finding confirmed a much debated theory about the geologic history of Hawaii. According to the theory, huge chunks of the islands collapsed into the ocean in ancient, catastrophic landslides. Current evidence suggests that the southeastern coast of the island of Hawaii may be next to go.

Multibeam sonars employ many focused beams of sound waves to chart topography. Based on the time it takes for the sound to return and on the geometry of the reflected beams, a computer can produce simulated 3-D images of volcanoes, ridges, trenches, valleys and other formations. Scientists are now mapping large portions of the ridge system. They hope to learn more about such fundamental processes as how magma is delivered to the ridges and what happens when it gets there.

Although mapping efforts are focused on the ridge system, the cluster of volcanoes discovered off the Mid-Pacific Rise suggests that much awaits discovery in other areas as well, according to Macdonald. Writing in *GSA Today*, published by the Geological Society of America, he said, "We hope to catch up, before the end of the milennium, with the successful mission to map the surface of Venus."

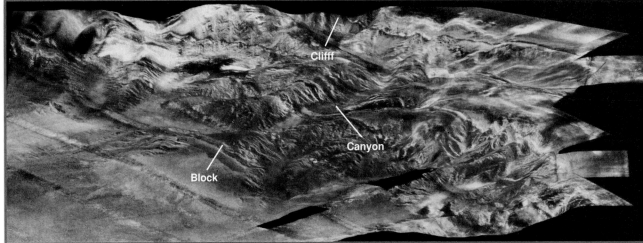

In 1983, the United States claimed sovereign rights within an Exclusive Economic Zone covering more than 3 million square nautical miles off its coasts. As part of an effort to characterize the geology within the zone, the U.S. Geological Survey turned to a sidescan sonar system called GLORIA.

The system emits sound waves down to the seafloor and analyzes the intensity of returning waves to produce black-and-white images of swaths of the ocean bottom. Hard, rocky regions, which scatter much of the sonar signal back to GLORIA, appear as bright areas. Softer sediments, which absorb much of the signal, are darker. When pieced together in a mosaic, the swaths present a reconnaissance view of the seafloor that resembles an aerial photograph. Scientists can electronically combine a mosaic with topographic information to produce a simulated 3-D view like those on this page.

Among GLORIA's finds in the Exclusive Economic Zone was a sinuous channel on the floor of the Gulf of Mexico (right) that is remarkably similar to rivers on land. Marine geologists believe the channel carries mud, sand and coarse debris. In one segment, a meander has formed into a loop that looks like an oxbow lake, a feature often associated with meandering rivers. (The narrow bands in the mosaic, which are spaced 18.6 miles apart, represent areas not imaged by GLORIA because they were directly under the track of the ship and thus out of reach of the sonar signal.)

A mosaic of GLORIA images from the Bering Sea off the coast of Alaska (above) reveals structural and textural details of the seafloor in and around 90-mile long Zhumchug Canyon, the world's largest known submarine canyon. USGS geologists believe that a large block at the end of Zhumchug may have broken free from the cliff in the background of the image, triggering vigorous erosion that led to the formation of the canyon.

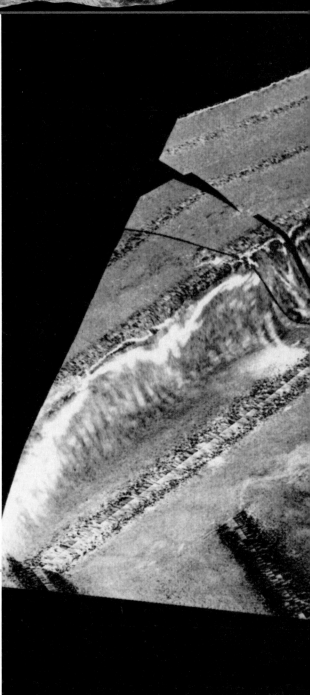

Multibeam sonar systems are revealing the detailed topography of the midocean ridge system, where the seafloor spreads apart and new crust is made. Along the East Pacific Rise, a portion of the ridge under about 10,000 feet of water off South America, the seafloor is spreading as fast as six inches per year. In 1991, scientists produced a 3-D view of a section of the rise near 8 degrees South latitude (large image). The slender, central part of the rise, called the axial high, is flanked by lower parallel ridges and seamounts.

In the image, which depicts an area three and a half times the size of New Jersey, the axial high is about 12 miles wide and ranges in height from about 600 to 1,300 feet. The seafloor to the left (west) is on the Pacific tectonic plate, while seafloor to the right is on the Nazca plate. Toward the south, the axial high curves into the Wilkes transform fault. Adjacent to the curve is an area of disturbed seafloor where small ridges are sheared and rotated. According to Jim Cochran of Columbia University's Lamont-Doherty Earth Observatory, the rapid rate of seafloor spreading may have seperated the crust here, forming an extremely small plate he calls a "nanoplate." To the north, the axial high is offset in a feature called an overlapping spreading center, which multibeam sonars have shown to be common on fast-spreading ridges. A view of the East Pacific Rise at about 9 degrees North latitude (top) shows an overlapping spreading center closeup.

The topography of slow-spreading portions of the ridge is more rugged. A view looking south along the Mid-Atlantic Ridge at about 31 degrees South latitude reveals a 12-mile-wide central valley studded with volcanic cones. The valley is bounded by mountains about 6,500 feet high. This axial rift valley is intersected at its southern end by the Cox transform fault.

According to Ken Macdonald of U.C. Santa Barbara, differences in magma supply account for the topographic differences. Fast spreaders steadily receive a robust supply of magma from below. Lava erupts from fissures along the crest of the rise, making new crust. The elevation of the axial high is a result of a bowing up of the crust caused by the magma that is pushing up from below. At slow-spreading ridges, where seafloor separates at about one inch per year, magma is supplied episodically from many discrete sources below the rift valley. The mountains lining the valley are volcanic constructions built by lava flows.

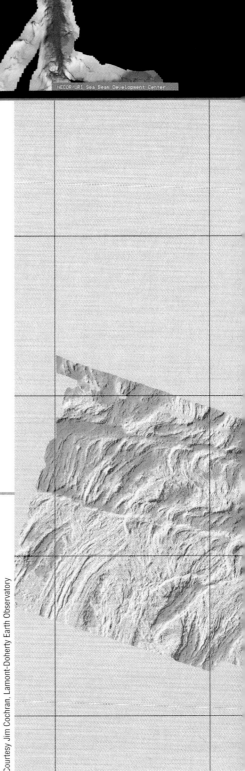

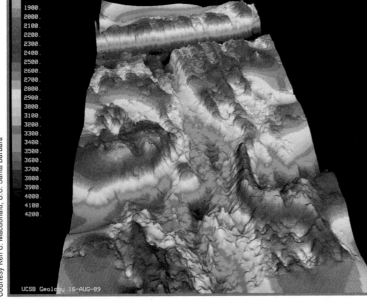

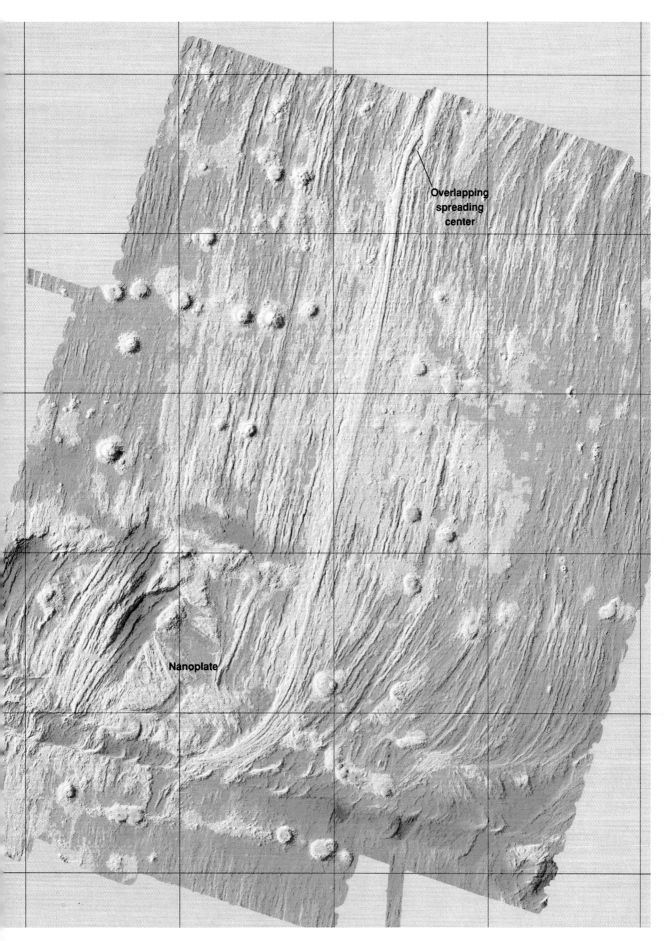

Field Trip:
Anatomy of a Mountain

By Berkeley Chew

Early snow drapes the landscape in the Sneffels Range, part of the San Juan Mountains of Colorado. Inset: The Animas River meanders through a valley carved by a glacier just north of Durango. Rocks on the mountainsides record 300 million years of geologic history.

Range

Berkeley Chew

Marc Muench

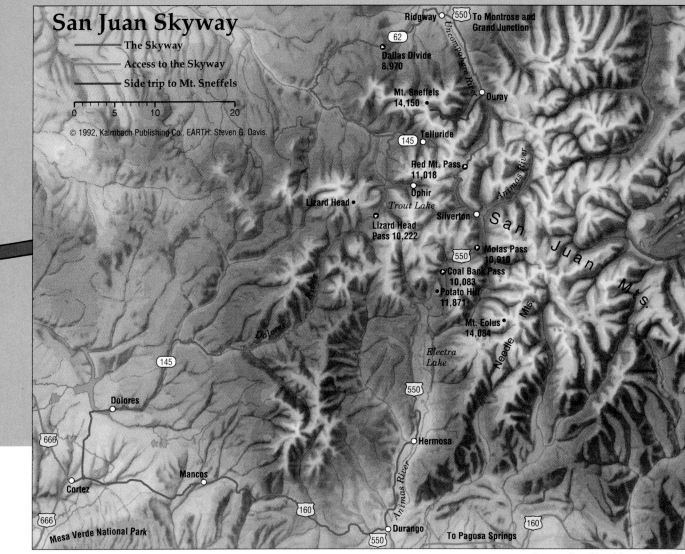

The Rocky Mountains of Colorado are a kind of geologic reincarnation, the latest of three mountain ranges that have stood in this part of the North American continent over the last 320 million years. The forces that created mountains here, wore them flat and created them again, have included collisions between giant pieces of the Earth's outer shell, the ebb and flow of seas, swamps and deserts, volcanic eruptions, and erosion by water and glaciers.

Nowhere in the Colorado Rockies is evidence of these geologic forces more spectacularly apparent than along the San Juan Skyway, a 236-mile loop of highways through the San Juan Mountains in the southwestern part of the state. A drive along the Skyway offers an unparalleled opportunity to learn about the geologic history of the Rockies while experiencing some of the finest mountain landscapes of North America.

You can begin anywhere along the loop, but the town of Durango is a good starting point. Beginning on U.S. Route 550 heading north out of town, the Skyway traverses a landscape of serrated peaks and deep canyons. It then makes a thrilling descent into the historic mining town of Silverton and into a region that was shaped by stupendous volcanism. Pressing farther north, the Skyway runs through Ouray and Ridgway, then curves southwest along other highways (see the map above for directions), passing among more high peaks. From the town of Cortez the Skyway heads east, completing the loop back in Durango.

Heading north out of Durango into the high country is a journey back in geologic time: from the U-shaped Animas valley, carved by a glacier 10,000 years ago, to the glittering high peaks, whose rocks formed when the foundations of this part of the continent were first being laid more than a billion years ago. Cobbling together the geologic history backwards from the features visible along the road is easier if you first understand the story's broad outlines.

About 1.8 billion years ago, today's Rocky Mountain region was more akin to Indonesia than Colorado. Sediment and volcanic debris accumulated in thick rock layers on and around a chain of islands that lay off the shore of a continent whose southern edge was in what now is Wyoming. Movement of tectonic plates, the

The steplike bands of rock in the open lands near Molas Pass are composed of limestone and shale that are believed to have formed in a large, partially flooded intermontane valley.

The Animas River flows south through a gorge leading toward and through the heart of the Needle Mountains. The Skyway is the road that passes through the colorful aspen in the foreground.

giant pieces of crust that comprise our planet's outer shell, carried these rocks eight to 10 miles down into the Earth's interior. Heat and pressure metamorphosed them into schist and gneiss. Erosion exposed these metamorphic rocks about 100 million years later. Today, geologists call them the "basement rocks," because all others have been deposited on top of them.

Approximately 1.4 billion years ago, the schist and gneiss were invaded from below by hot magma, which crystallized to form granite intrusions.

About 320 million years ago tectonic movements brought the southern part of North America into contact with northern South America. This slow-motion "collision" (which took place at about the rate a fingernail grows) raised and faulted large blocks of land, creating the Ancestral Rockies — the region's first alpine incarnation. The area of today's San Juan Mountains lay just outside the main area of uplift, and over the course of many million years became buried in thousands of feet of sediments shed from the mountains as they eroded flat.

Mountains reappeared during a second uplift called the Laramide Orogeny. Geologists believe this was initiated by subduction, in which movement of the tectonic plates drives sea floor under the edge of a continental landmass. About 70 million years ago, subduction of Pacific Ocean sea floor beneath the western margin of the North American continent quickened. The sea floor melted partially as it plunged into Earth's interior, heating the continental crust above it. From the coast to present-day Colorado, the continental crust buckled and shortened like an accordion. In the process, blocks of basement rocks and overlying sedimentary layers were uplifted and other blocks dropped.

A period of volcanism began about 35 million years ago, as the Laramide mountains were being eroded to a rolling plain. Volcanoes erupted thousands of feet of lava and volcanic ash on the land. Then huge depressions called calderas formed. They were ringed by other volcanoes that spewed hundreds of cubic miles of hot ash, which settled in and around the calderas and solidified, forming brightly colored layers of tuff. Today, the most impressive evidence of this geologic violence is seen in the San Juans. When the violence was over, the mountains started to wear away, and they eventually became half-buried by rubble eroded from their heights.

The third and final uplift began 26 million years ago, when the subduction that had caused shortening of the crust during the Laramide Orogeny began having the opposite effect: much of the crust of western North America began to stretch. Volcanic and underlying formations left after erosion of the Laramide mountains were broken by faulting, and the whole region was uplifted. Water and glaciers went to

Hidden within high valleys of the volcanic mountains north of Silverton are the ruins of old mining camps, including this aerial tramway that once carried ore buckets.

work on the highlands. The basement rocks in the cores of the Laramide uplifts, the layers of hard sedimentary rocks on their flanks, and the intrusive bodies of granite were more resistant to erosion and thus formed high peaks. Less resistant rocks were carved out, forming valleys.

Evidence of glaciation, one of the last acts in the creation of today's landscape, is ubiquitous in the Animas valley just north of Durango. A river of grinding ice scoured sheer cliffs into the mountains rimming the valley. The Animas River meanders back and forth through the valley and down the shallow grade left by the glacier, as if it doesn't know which way to go.

Stop a half mile north of Durango and look at the long, forest-covered mountains on either side of the valley. They rise about 2,500 feet above the valley floor and are capped by a conspicuous line of 50-foot-high buff-colored cliffs. The rocks in these cliffs formed as a sea advanced into the region about 85 million years ago, long after the Ancestral Rockies uplift but before the Laramide Orogeny. The sand from beaches and bars along the shores of this sea coalesced into widespread deposits of white sandstone, today known as the Dakota Sandstone. The sandstone was uplifted into its present position during the Laramide Orogeny and the most recent episode of mountain building.

Greenish rocks on a sparsely vegetated slope under the Dakota Sandstone are shales of the 150-million-year-old Morrison Formation. This formation was deposited toward the end of the long period of erosion that wore down the Ancestral Rockies. Lazy, muddy rivers flowed across a large, swampy plain inhabited by dinosaurs. These rivers laid down layer upon layer of silt and mud. The resulting shales are today a treasure-trove of dinosaur fossils.

Under the Morrison Formation is a 50-foot-high band of white cliffs. These are composed of sandstones of the Entrada Formation. (In the southwestern deserts, the Entrada is orange.) The sandstone formed about 155 million years ago, during the later stages of the erosion of the Ancestral Rockies, when the region was a dune-filled desert.

Below the white cliffs is a thousand-foot-high slope of dark red ledges and cliffs. This is the 280-million-year-old Cutler Formation, which formed when the Ancestral Rockies were at their highest. Rocks stripped off the peaks were carried by fast-moving streams and deposited in and around the mountains, forming thick layers of sandstone and conglomerate.

About 12 miles north of Durango the highway begins to climb up the western side of the Animas valley, and you soon lose sight of the Animas River below. To the northeast of Electra Lake (which is not visible from the road), the Animas River runs out of a gorge through a rugged country of high peaks. The narrow-guage Durango and Silverton railway follows this

Evidence of one quarter of the Earth's geologic history has vanished between the vertical layers of 1.4-billion-year-old quartzite in Box Canyon near Ouray and the 355-million-year-old limestone layers that lay on top — a stunning unconformity.

canyon on its way to Silverton. The Skyway heads straight north.

When you come to the Needles Country Store, you'll see three sharp peaks of the Needle Mountains towering to the east. The farthest is Mt. Eolus, 14,084 feet high. These high peaks are made of 1.4-billion-year-old granite, a spectacular example of the intrusions that invaded the metamorphic basement rocks. The walls of the Animas Gorge far below the peaks are 1.8-billion-year-old basement rock itself.

Route 550 passes the Purgatory Ski Resort, climbs northeast and then switches back sharply before turning generally northeast for the final pull up to 10,640-foot Coal Bank Pass. Pull over on the wide shoulder before you get to the pass, not just to smell the balsam of the subalpine forest, but also to look closely at basement rock in the roadcut, which is dug into the flank of Potato Hill. You should look for dark-orange, banded rocks streaked with quartz. Geologists call it "Twilight Gneiss."

At the crest of Coal Bank Pass the scene changes abruptly. To the east and north spread open lands whose slopes appear to have numerous long steps cut into them. The steps are alternating layers of limestone and shale of the Hermosa Formation. These rocks formed during the uplift of the Ancestral Rockies in a large intermontane valley that was partially filled with water.

Farther on, the highway passes over Molas Pass, plunging downhill and further back in geologic time toward Silverton. You pass through a layer of bright-red material that looks like a kind of muddy soil. This is the Molas formation, and it is, literally, a fossil soil. It accumulated about 320 million years ago, when this area of the continent was a wet jungle punctuated by limestone towers and sinkholes — a karst landscape similar to areas of Venezuela and Southeast Asia. The Molas soil, which never hardened into rock, accumulated atop the limestone — the next sequence of rock layers, called the Leadville Limestone, that you encounter on the way down to Silverton.

Before reaching Silverton you pass through ancient sedimentary formations laid down directly atop the basement rocks before the uplift of the Ancestral Rockies. And then you enter a completely different, younger world: the rocks of the mountainous terrain to the east, north and west of Silverton were produced by volcanism that began 35 million years ago, toward the end of the Laramide Orogeny. The volcanism ended about 26 million years ago, just prior to the last period of uplift. Toward the end of this period, the land Silverton sits on and a wide region to its north sank by as much as 3,000 feet into an empty volcanic cauldron. This huge collapse created a caldera 10 miles wide, as well as a roughly circular system of cracks in the rocks around its periphery. It was in these cracks that lodes containing gold, silver, lead and tin formed.

In the 1870s miners founded Silverton and other towns and

JANUARY 1993 **69**

This view southwest from atop Mt. Sneffels, elevation 14,150 feet, takes in a rock glacier, which flows down from the nearby mountain crest, an aspen-covered valley in the distance, and the San Miguel Range beyond.

The 400-foot-high pinnacle of Lizard Head, located near Telluride, is made of volcanic material. Lizard Head can be seen from the Skyway at Lizard Head Pass.

camps around the rim of the caldera. Today, the mountains are literally riddled with their now disused tunnels. Old prospect holes, overhead tramways, mine portals and mine dumps remain in the high valleys, slowly falling into ruin.

From Silverton, Route 550 curves west and north along Mineral Creek, which is the western edge of what was the caldera. The eastern edge is marked by the upper Animas River. All along Mineral Creek up to Red Mountain Pass you will see hills and mountains made of brilliant orange and yellow rocks: remnants of the tuff formed from the ash that spewed from the caldera 26 million years ago.

After the pass, the Skyway makes a dizzying descent into the town of Ouray. Almost as dizzying as the descent is the unusual juxtaposition of rock formations in Box Canyon, located a short distance off the road just before you get into Ouray. Turn left at the sign showing the way to the canyon. A short path takes you past spectacular Box Canyon Falls. Just beyond the falls you will see layers of rock stacked vertically like books on a shelf. These are overlain by horizontal layers, forming one of the finest examples anywhere of what geologists call an "unconformity."

The vertical layers are composed of quartzites that were originally laid down horizontally as sandstone 1.7 billion years ago atop older layers of basement rock. Tectonic forces generated by subduction under the margin of the continent — a continuation of the subduction that led to the metamorphosis of the basement rocks — tipped the sandstone layers on end and metamorphosed them slightly, forming quartzite. The horizontal rock layers seen today atop the vertical quartzites are 355-million-year-old limestones. Rocks representing about 1.3 billion years of geologic history — one quarter of the age of the Earth — are missing between the vertical and horizontal layers. They were eroded away before the horizontal limestone layers were deposited.

Hikers may want to consider a climb on 14,150-foot Mt. Sneffels, located about seven miles west of Ouray. Mt. Sneffels is made of hard igneous rock — probably the remains of magma that solidified in a passageway that was feeding a volcano. Flanking mountains and ridgelines are made of layers of tuff. The climb is not for novices, but the fit and experienced can do it in a day from the Yankee Boy Basin trailhead. Access is provided by Ouray County Road 361, which intersects the Skyway just south of Ouray. Those interested in climbing Mt. Sneffels should plan carefully with the help of a guide-

The Silverton-Durango narrow guage railway steams through the narrow Animas Gorge, whose walls are made up of 1.8 billion year old gneiss. The railroad offers a different way to learn something of the geologic history of the San Juan Mountains on a thrilling trip between Durango and Silverton.

book such as *Colorado's Fourteeners*, by Gerry Roach.

About four miles north of Ouray Route 550 follows the broad, U-shaped Uncompahgre valley that is a mirror image of the Animas valley near Durango. The geology is nearly the same here, from the sheer mountainsides down to the gently meandering river. About 12 miles north of Ouray you come to Ridgway. Turn left on Route 62. From here the road climbs toward Dallas Divide. The panorama to the south of Mt. Sneffels and flanking snow-covered peaks is renowned — one of the most beautiful spots in America. When the aspen turn bright yellow-orange in autumn, the view is quintessential Colorado.

At Dallas Divide you have not completed even the first half of the Skyway loop, yet you have seen evidence of all the major phases in the geologic history of the San Juans and Colorado Rocky Mountains. But much spectacular scenery and geology lie ahead. In the town of Telluride, located within the region shaped by volcanism, you can see Colorado's highest waterfalls, Bridal Veil Falls. Farther along is Trout Lake. Come here around sunset, when a mountain wall behind the lake glows as if it is on fire. A little farther on, just beyond Lizard Head Pass, pull over for a view of Lizard Head, a 400-foot sheer rock tower formed from volcanic material.

From here, the Skyway leads southwest along the Dolores River to the town of Dolores. A tableland lies to the south: Mesa Verde, or green mesa, whose rainy top supported the cornfields of ancient American Indians. Their cliff dwellings are magnificently preserved in Mesa Verde National Park.

The last part of the loop starts just east of Cortez. Route 160 takes you east through the town of Mancos and over the southern shoulder of the little-visited La Plata Mountains. As you sweep down toward Durango, pull over for a vista of the San Juan Mountains to the east and north.

If you try hard, you may be able to imagine all those layers of rock forming in seas, floodplains, swamps and deserts. Perhaps you can picture the Ancestral Rockies rising up and being worn down, only to be replaced by the Laramide mountains. And maybe your mind's eye can conjure a vivid image of volcanoes spewing lava and ash over the eroded remnants of the Laramide ranges, and of a third mountain range rising up and being incised by rivers and glaciers to create the landscape you see in the distance. With more imagination, you may even see the geologic future, as the mountains erode flat only to be reincarnated again. ⊕

When not wandering the mountains, Berkeley Chew writes fiction and nonfiction. He lives in Durango, Colorado.

The editors of Earth *magazine wish to thank John C. Reed, Jr., a geologist with the U.S. Geological Survey in Denver, for his help in preparing this article.*

Lessons from Landers

The huge earthquake that rumbled out of the Mojave Desert near Landers, California, last June was the largest to strike the state in 40 years and the fourth largest in the state's history. The unusual quake provided geologists with new evidence that California may someday tear away from North America and drift toward Alaska.

By Richard Monastersky

The Landers quake caused limited damage because it struck a sparsely populated area. But the ground ruptures it caused may topple basic assumptions about how quakes work.

Few people can forget the arresting images of Northern California after the Loma Prieta earthquake of October 17, 1989. With the nation watching on television, burning buildings in San Francisco set the night sky aglow with eerie orange hues. In Oakland, rescue crews worked around the clock to reach drivers trapped within a section of Interstate Highway 880 that had collapsed like a house of cards. Whole cities were shut down by a disaster that caused more than $60 billion in damage and took 63 lives.

In human terms the quake that rocked Southern California last June 28 carried much less drama. This temblor emanated from the isolated Mojave Desert, near the town of Landers, and caused relatively little damage when compared with the Loma Prieta shock.

Nonetheless, the Landers quake was the largest jolt to shake California in four decades. Measuring a magnitude of 7.5 on the Richter scale, it packed about five times the energy of Loma Prieta. And because this powerful quake hit within a region covered by seismometers, the Landers quake has offered researchers unprecedented opportunities for discovery.

"I think the Landers event will go down in history as one of the most important earthquakes from a scientific standpoint," says Roy Dokka, a geologist from Louisiana State University who studies the Mojave Desert.

Many news reports immediately following the earthquake emphasized that the Landers shock

MARCH 1993 **73**

increased the risk of the Big One — the huge earthquake that seismologists have warned may soon strike Southern California. But the impact of the Landers earthquake extends far beyond the immediate vicinity of its epicenter. As scientists delve into the details of the recent event, they are learning new lessons about how earthquakes work in general. The desert quake has also raised some intriguing questions about the geologic future of California.

Hopscotch quakes

Landers was not only powerful — it was bizarre. In most earthquakes, the land tears along a single fault. But in this quake, the rupture hopscotched from fault to fault, growing more powerful with each successive jump.

From its birthplace just south of Landers, the rupture raced almost due north for about 12 miles, took a sharp curve to the northwest, and then continued for another 31 miles. In the end, the energy released by the Landers quake caused movement along perhaps five different faults, according to geologists who studied surface scars created by the jolt.

Seismologists studying how the earthquake behaved below the surface noticed this strange pattern too. Although the Landers quake felt like a single one to humans, records of the seismic waves generated by the quake reveal that it actually was a pair of successive shocks. The first set of vibrations to arrive at seismic stations came from a fault in the south that was pointing north. About ten seconds later, a larger set of seismic waves emanated from a set of faults located farther north; these faults were pointing northwest.

Few earthquakes make such drastic bends and turns, seismologists and geologists say. In fact, large breaks and bends in faults are usually thought to stop a rupture from spreading. "Normally people think that the rupture is limited within a very straight section. But if you look at this [earthquake], it's very obvious that it doesn't work that way all the time," says Hiroo Kanamori, director of the Seismological Laboratory at the California Institute of Technology.

Scientists have previously witnessed this type of behavior during large earthquakes in Asia, but this is the first time a major shock in California has so clearly demonstrated the pattern, says Lucile Jones, a seismologist with the U.S. Geological Society in Pasadena.

The Landers example raises disturbing questions about a common practice used in estimating the future hazard from faults. Geologists often study and measure fault breaks to predict the likely size of future quakes along the fault. "It's common practice to find a fault, see how long it is and use that to estimate the maximum credible earthquake," says Jones. But the Landers quake shows that ruptures can jump major hurdles, suggesting these estimates may have little meaning.

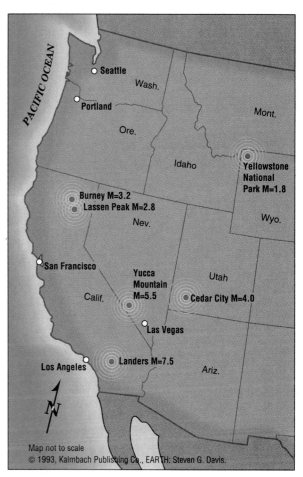

On the same day the Landers quake struck Southern California, numerous other earthquakes rattled localities far from the epicenter. A swarm of tiny jolts hit as far off as Yellowstone National Park (left) 744 miles away. A magnitude 2.8 shock rumbled near Lassen Peak (above) just minutes after the Landers quake. The map at right pinpoints several of the quakes apparently triggered by the Landers shock. Until the Landers quake, seismologists did not believe that one earthquake could trigger activity hundreds of miles away.

Distant quakes

Landers surprised researchers in other ways as well. The quake's effect on distant faults throughout the western United States was even more unusual than its complex, jumping movements. The same day the Landers earthquake struck Southern California, a swarm of tiny jolts erupted beneath Yellowstone National Park, 744 miles away. Other jolts occurred in the far north of California and in western Idaho, Utah, and Nevada.

In the past, most seismologists would have scoffed at suggestions that one quake could trigger activity at such great distances. Indeed, researchers have previously written off earthquakes that occurred on the same day in distant locations as sheer coincidence. After all, hundreds of minor earthquakes occur in California each week.

But that was before the Landers quake shook up conventional seismological wisdom. "It's something that nobody would have expected," says James Dieterich, a geophysicist at the USGS in Menlo Park, California.

On the day of the Landers quake, small tremors broke out at various sites around the state of California, particularly along the eastern Sierra Nevada Mountains and beneath Lassen Peak and Mount Shasta — two volcanoes in the north. A magnitude 2.8 quake struck near Lassen Peak just minutes after the Landers quake. More than a day later, a magnitude 3.2 quake hit near the town of Burney, located about halfway between Mount Shasta and Lassen Peak. In Utah, the Cedar City area suffered a swarm of tremors, some as large as a magnitude 4.0. An even larger shock measuring 5.5 rattled southern Nevada near Yucca Mountain, where the federal government is considering building an underground nuclear waste repository. By contrast, the earthquakes beneath Yellowstone were only as large as a magnitude 1.8 — too small for people to feel.

In this situation, the relationship between the large quake and accompanying tremors was simply too obvious to ignore. So many areas became active — as if someone had flicked on a switch. Tremors started in regions hundreds of miles away from Southern California mere seconds after the Landers quake.

Paul Reasenberg, a seismologist with the USGS, thinks strong earthquakes must have triggered distant seismic activity in the past. But seismologists didn't notice the pattern before the Landers quake because California and other states had only recently set up networks of sensitive seismometers capable of detecting the numerous small earthquakes. When the last quake of Landers' magnitude hit, the networks were not in place.

Following the Landers quake, USGS seismologist William Ellsworth searched earthquake records for signs of distant triggering in the past. The great San Francisco earthquake of 1906 turned up some tantalizing possibilities. That disaster, which leveled

The great San Francisco earthquake of 1906 skewed some buildings and leveled many others. Nine other earthquakes struck California and Nevada within two days of the disaster. Many geologists believe this supports the theory that large quakes like Landers can trigger distant temblors.

much of the city and sparked disastrous fires, occurred at 5:30 on the morning of April 18. Within the next two days, nine sizable earthquakes shook up California and Nevada, Ellsworth found. One of these quakes had an estimated magnitude of 6.2 and struck clear across the state at the far southern end.

Indeed, the pattern in 1906 resembles the activity following the Landers shock, Ellsworth argues. Much of the activity occurred to the east of the Sierra Nevada Mountains, along a belt that has numerous faults and potentially active volcanoes.

No one has satisfactorily explained the triggering phenomenon, however. Scientists don't know how one earthquake can set off others hundreds of miles away, but they've developed a few theories.

Seismologists have long known that an earthquake on one fault can raise the stress on other faults, inducing them to produce jolts of their own. In fact, stress from the Landers quake triggered a magnitude 6.6 aftershock beneath the nearby town of Big Bear three hours after the main shock. But these so-called "static" stress changes have little effect far from the main earthquake. Robert W. Simpson of the USGS has calculated that the Landers shock would have induced static stress changes at Mount Shasta and Lassen Peak that were far weaker than the stresses created each day by the tidal pull of the sun and moon. Researchers say it's unlikely that such puny changes could have triggered the sympathetic vibrations in this case.

The tremors that accompanied the Landers quake were more likely caused by traveling vibrations, the stresses that the seismic waves create as they pass through a region. Unlike the static stress changes, the stresses associated with seismic waves last only a few moments. But the traveling vibrations remain stronger than the static stresses at a distance from the epicenter.

One way traveling vibrations could trigger bursts is by altering the friction on faults they pass. You might think of this concept as the "stuck window model."

Like a stuck window, pieces of land on either side of a fault are held together by frictional forces that prevent them from sliding past each other smoothly. If a person bears down on a stuck window with increasing force, the window will eventually slam shut with a bang (and probably some broken glass). Similarly, sooner or later stresses in the Earth grow until they overcome the frictional forces. Then one piece of land jerks forward, up, down or sideways. This is an earthquake.

Everyone knows that a sharp knock on the window frame sometimes does the trick too, loosening the window enough for it to close freely. In much the same way, the seismic vibrations from a larger earthquake could somehow lower the frictional forces on a distant fault, causing the stuck pieces of land to slip. But seismic waves may or may not remain forceful enough to lower the friction appreciably on a distant fault. In other words, says Dieterich, by the time the vibrations from Landers reached Mount

During the Landers earthquake, pieces of land on either side of the ground rupture slid past each other, offsetting fences around the property above. The Landers quake, four earlier temblors, and a strong 1908 quake in Death Valley all line up (map, right). This suggests to geophysicist Amos Nur that an incipient fault may supplant the San Andreas as the boundary between the Pacific and North American plates. Geologist Roy Dokka argues that the quakes occurred not on an incipient fault but within the Eastern California Shear Zone, but he agrees that a new plate boundary may be forming here.

Shasta, did they have the force of a hammer hitting the window — or of a cotton ball?

Some scientists think the behavior of a bottle of cola might better explain the triggering effect. Many of the seismic bursts following the Landers quake occurred in volcanic regions that overlie chambers of molten rock or magma deep in the Earth. As the seismic waves passed through these magma chambers, they may have caused dissolved gases to come out of solution, just as shaking a bottle of cola will cause it to foam. The bubbles of volcanic gas may then have pushed against the walls of the magma chamber, creating small earthquakes.

Researchers are exploring other possible explanations for the triggered earthquakes as well. But so far, no one idea has emerged as a clear winner. "We're still searching. It's going to take a while," says USGS seismologist David Hill.

A new fault is born?

While seismologists try to solve the triggered-earthquake problem, experts in plate tectonics are attempting to fit the Landers shock into the broader picture of North American geology. One group of geoscientists thinks the recent earthquake may be part of a new fault being born in the desert, a fault that may some day play a major role in shaping the western edge of the continent.

This notion emerged out of plate tectonics, the theory that explains how Earth's surface continually reshapes itself. Like a cracked eggshell, the planet's outer layer is broken into more than a dozen large pieces called tectonic plates. These plates float like buoyant rafts atop a hot, semimolten layer in Earth's mantle. Averaging about 60 miles thick, they constantly shift position, driven by moving currents of pliant rock inside the mantle.

Where the plates crash together they form mountain ranges or deep trenches. Where they slide past each other they create major rips in the Earth's shell; the San Andreas Fault is one such tear. These battles between the plates generate the stresses that spark most earthquakes around the world. Over decades and centuries, the stresses build until they literally break the Earth's crust, either along new faults or old ones that have been resting quietly.

The San Andreas Fault represents the boundary between the plate underneath the Pacific Ocean and the plate carrying North America. The movement of the Pacific Plate relative to the North American Plate is squeezing the Earth's crust in the Mojave Desert.

This stress building in the Mojave would be relieved if the faults in the desert were oriented in a north-south direction. But curiously, most faults in this region point to the northwest — a direction wholly unsuited for absorbing the stress in the Mojave, says Amos Nur, a geophysicist at Stanford University.

MARCH 1993 **77**

Three years ago, Nur and his colleagues suggested that these awkwardly directed faults did point in the most favorable direction at one time. Since then, however, the faults rotated to their present orientation. Think of what happens to books on a shelf when you move the bookend. Without the bookend's support, each book slides against its neighbors, and the entire line of volumes leans over. Viewed from their bindings, the books appear to rotate.

According to Nur's theory, the faults in the Mojave Desert are akin to the border between two books: The faults are parallel and they allow one block of crust to slide against an adjacent block. At first the faults pointed in a north-northwest direction, but over 6 million years they all rotated by 15 to 20 degrees until they pointed almost due northwest.

Nur and his colleagues noted that four Mojave quakes during the last few decades did not fall on the expected northwest trending faults. Instead, these anomalous jolts occurred on faults that point about 10 degrees to the west of north, precisely the direction that would best accommodate the stress in the region. Nur's group suggested that a new set of faults was forming that would eventually replace the older generation.

The Landers earthquake supports that hypothesis because it started on a fault pointing in the same direction as the four earlier anomalous earthquakes, says Nur. When plotted on a map, Landers fits on the same line as these quakes, including one that occurred on April 22, 1992, in Joshua Tree National Monument. If this line represents a fault in formation, it would be sizable, measuring about 60 miles long.

Replacing the San Andreas

Nur's group thinks the fault may be even longer, though. When the researchers project the line farther toward the north, it runs right by the site of a strong earthquake that struck Death Valley in 1908. If the 1908 tremor also lies on the same incipient fault as the Landers quake, the Joshua Tree quake, and three other major earthquakes that occurred in recent years, the entire fault would stretch over 155 miles, about a quarter of the span of the San Andreas. In that case, we may be watching the birth of a young fault that will challenge the San Andreas as the active boundary between the Pacific and North American tectonic plates, says Nur.

The venerable San Andreas is open to competition because it has a major flaw that makes it a less-than-ideal boundary between the Pacific and North American plates. The fault runs a mostly straight course from the northern part of California to the southern end of the state. But when it nears Los Angeles, the fault is warped by a big bend — a shape that makes it difficult to accommodate the movement of the plates.

Over the years, many geoscientists have sug-

A long rupture in the Mojave (left) opened during the Landers quake. Land on either side moved laterally and horizontally (right). Lines marking spaces in a church parking lot were offset by the quake (below).

gested that the plate boundary would eventually shift eastward because of this major bend. If Nur's new fault is indeed forming in the Mojave, this youngster would be ideally suited to absorb stress that the San Andreas cannot. Perhaps movement of land along the San Andreas will slow down as the new fault grows more active, eventually causing the plate boundary to jump to the east.

Of course, there are many possible scenarios, and some geologists have roundly criticized Nur's theory about a newly forming fault. "I have very good reason to believe that what he's saying is not correct," says Louisiana State's Dokka, who has spent 17 years studying the geology of the Mojave.

Dokka contends that no fault currently exists in the position that Nur suggests. Instead of a single fault, he argues, there is a broad belt of faults called the Eastern California Shear Zone that play a critical role in the plate tectonics of western North America.

Indeed, the plate tectonics of the American West is actually more complex than most people paint it. The San Andreas Fault is often called the boundary between the two plates, but such a statement is a bit oversimplified. The plate boundary, in reality, also includes other faults across a wide region.

About 70 percent of the movement between the Pacific and North American plates occurs along the San Andreas. Geologists have long thought that half of the remaining movement occurs along faults in the Basin and Range province. But they are unsure where the remaining movement occurs.

Some think underwater faults west of Los Angeles do the job. However, Dokka and his colleagues recently discovered that movement along faults in the Eastern California Shear Zone accounts for the motion left over from the San Andreas and the Basin and Range.

In Dokka's view, the Landers earthquake provides dramatic proof that the Eastern California Shear Zone is indeed absorbing part of the plate motion. The Landers quake does not represent the birth of a new fault; instead, it simply shows how existing desert faults have been working for at least 6 million years.

Looking into the future, though, Dokka agrees with the general conclusion that eventually the plate boundary may shift eastward, splitting off most of California from the rest of North America. That's what happened farther south about 5 million years ago when the plate boundary jumped to the east of Baja California, shearing this block off the Mexican mainland and creating the Gulf of California.

If the plate boundary does move to the east of the Sierra Nevada, that would tear off most of California and send it on a slow ride toward Alaska. But don't go selling any of that beach-front property in Southern California just yet. This process will take millions of years to work itself out. ⊕

Richard Monastersky is the earth sciences editor at Science News.

CAT scanning the

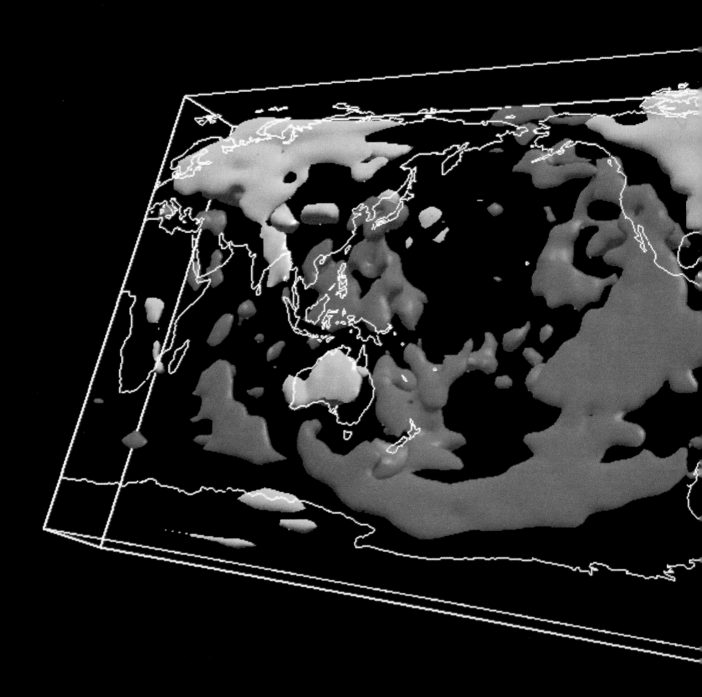

In the Earth's mantle, 60 to 450 miles underground, huge blobs of hot rock (red) rise toward the surface while cold blobs (blue) sink. The Pacific Ocean is ringed by hot blobs.

Earth

The latest in computer graphics enables geophysicists to visualize the insides of our planet.

by Jim Dawson

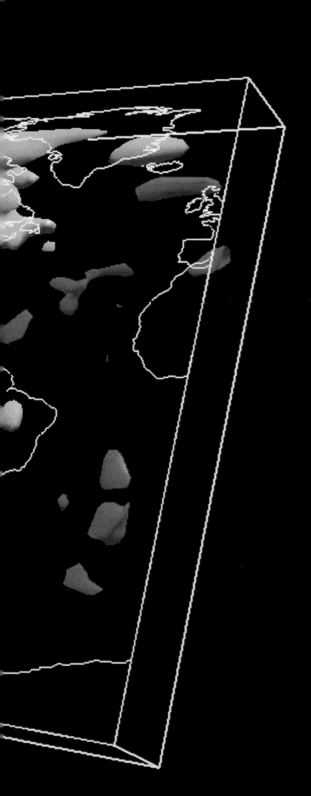

Producing vivid, three-dimensional pictures of the insides of the Earth was the last thing Paul Morin imagined he would be doing as a linguistics major at the University of Minnesota. The structure of language, not the mantle, was foremost in his mind. But the 26-year-old undergraduate, who took a job at the Minnesota Supercomputer Institute to help pay his college expenses, has devised a method for visualizing the invisible: a program that depicts on a computer screen the internal structure of the Earth.

Like a CAT scan that has detected a tumor deep inside a person's body, Morin's imaging technique has revealed a mammoth spike of cold rock suspended from the underside of the crust of North America. This spike may be the root cause of previously unexplained large magnitude earthquakes near New Madrid, Missouri. But Morin's method and similar techniques are helping scientists do much more than diagnose mysterious earthquakes. They are also helping them map tremendous currents of hot and cold rock in the mantle, currents that give rise to volcanoes and drive the movements of the tectonic plates. This information is helping other scientists improve the accuracy of computer models used to simulate — and thus understand — how the insides of our planet work.

As a computer programmer at the institute, Morin had been adapting specialized software to visualize Minnesota geophysicist David Yuen's ideas about the mantle, the 220-mile-thick region of rock between the Earth's core and its crust. Yuen is a mantle modeler; like an aircraft designer who uses a supercomputer to simulate how a new design will fly, he uses a computer to simulate how the mantle works. In both cases, the behavior of what is being simulated can be viewed on a computer screen: currents of air passing over a proposed aircraft's wings and fuselage; convection currents of rising hot rock and sinking cold rock in the mantle.

To improve how his computer program portrays these mantle convection currents, Yuen suggested that Morin attend a lecture by a visiting scientist, Toshiro Tanimoto of the University of California at Santa Cruz. Unlike Yuen's hypothetical simulations of the mantle, Tanimoto uses a technique called seismic tomography to observe it — albeit indirectly. He does this by analyzing how seismic waves generated by earthquakes move through the interior of the Earth. The end product is a crude map portraying the present configuration

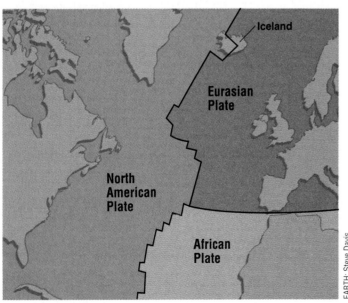

The North Atlantic Ocean is growing wider as the Eurasian and North American tectonic plates move away from each other. The plate boundary runs up the middle of the Atlantic and right through Iceland (map, left), where massive lava flows related to the rifting have been carved into spectacular landscapes by moving water (large photo). A computer image (above) reveals how hot mantle rock (red) rises to fill the gap opening between the plates.

of convection currents in the mantle.

As Morin watched Tanimoto's presentation he thought to himself: *terrific data, primitive images*. Like most seismic tomographers, Tanimoto was using two kinds of images: maps that show the temperature contours of horizontal slices of mantle and simple 3-D snapshots that show particular regions in perspective.

When the lecture was over, the linguistics student walked up to the well-known geophysicist and offered to transform the maps into images that could be interpreted much more easily — 3-D images that Tanimoto and his colleagues would be able to quickly and easily rotate on the computer screen. Tanimoto accepted the offer, and the next day he retrieved his data from California via telephone and turned it over to Morin. In a matter of hours Morin had adapted some of the programs he used in his work with Yuen to produce stunning maps of the mantle.

In the maps, large blue blobs represent regions of mantle rock that are colder than average, and red blobs represent regions that are hotter than average. (The exact average temperature of mantle rock is unknown, but scientists believe it is in the thousands of degrees Fahrenheit. They also believe that regions of "hot" and "cold" vary by about 300 degrees.) The blobs are connected in three-dimensional space, forming a complete picture of the temperature structure of the mantle. The findings of mantle modelers would suggest that the regions of hot mantle rock represented by the red blobs are rising and that the regions of cold rock represented by the blue blobs are sinking, thereby creating complex convection currents.

Joy Spurr/Bruce Coleman, Inc.

Morin says that when he showed his images to the researchers, the normally reserved Tanimoto became "quite excited," and Yuen "was bouncing off the walls." The two geophysicists were so excited because they had seen something complex and difficult to grasp quickly transformed into something almost instantly understandable. And they knew that Morin's visualization technique would help form a bridge between the explorational work of seismic tomographers and the hypothetical work of mantle modelers.

Geophysicists have been trying to visualize the Earth's interior for more than 70 years. Their main tools, seismic waves, are somewhat analogous to sound waves. As with sound, the speed of seismic waves depends on the nature of the materials through which they flow. Seismologists examine the time it takes for a wave to travel from an earthquake epicenter to a seismograph that may be thousands of miles away. In this way they can infer something about the rock that lies in between. Seismic waves pass through hotter rock more slowly than through colder rock. Thus, the "hot" (red) regions in Morin's maps represent areas of the mantle where seismic waves slowed down as they passed through the Earth. The "cold" (blue) regions represent areas where the waves sped up.

Early seismic-wave researchers discovered that the planet is divided into a shallow crust, a deep core (whose outer layer is molten), and in between, the solid region of the mantle. But they learned little else of the internal structure. It was not until the late 1970s that researchers had both a large enough set of earthquake data and a powerful enough set of mathematical techniques to get their first inkling of the deep Earth's structures in all three dimensions. Researchers began to call the technique seismic tomography, by analogy with the computerized axial tomography or CAT scan technique used in medical imaging. Seismic tomographers use the slowing of seismic waves in much the same way CAT scanners used the blocking of X-rays to produce images of internal structures.

At first they detected only the broadest outlines of the Earth's interior. By 1984 the smallest objects that could be seen lurking in the rocky deep were 1,500 miles across and 300 miles thick. Today, researchers can resolve objects 650 miles across and 15 miles thick, a substantial improvement but one that still leaves researchers hungry for more detail.

Morin's computer programs will not by themselves create that detail; to do so mainly requires collecting more seismic-wave data. But they do enable researchers to use the information they have in the most direct and intuitive way possible.

Researchers no longer have to rely on 2-D maps or on still snapshots. Morin's images are still confined to a two-dimensional computer screen, but just barely. The images on the computer screen can be rotated electronically in any direction, enabling researchers to see them from any angle and even watch them turn. The computers Morin uses are fast and powerful enough not only to produce instantly the view a geophysicist wants, but to show the image in motion as it turns to the desired position.

"It is the difference between seeing a coffee cup sitting on a table and being able to grab that coffee cup and manipulate it," Morin says.

Tanimoto's lab and Yuen's lab are still the only ones that regularly use Morin's interactive images. But geophysicist Adam Dziewonski and his colleagues at Harvard University, who also use seismic tomography to survey the mantle, have tried out the system and hope to acquire something similar as soon as they can get powerful enough computers. With such a system, Dziewonski says, "we essentially get a better perception of the actual distribution of the anomalies. You get a much better feeling for it."

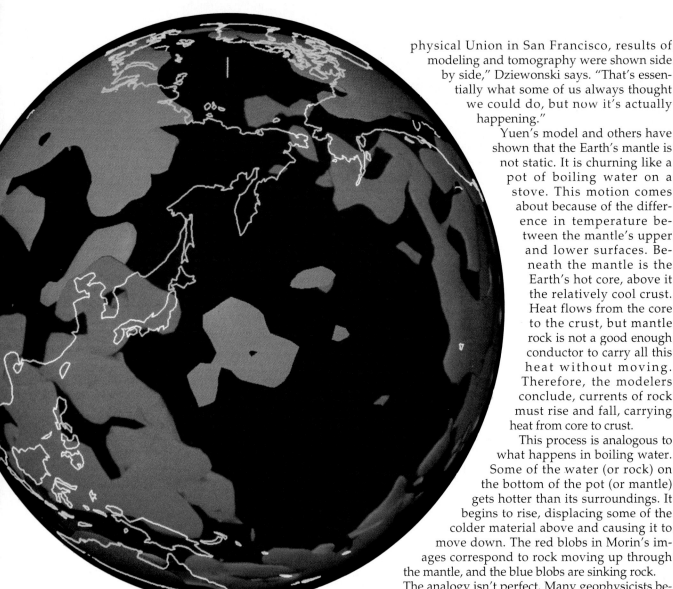

Courtesy Paul Morin

Computer images of the mantle have already helped geophysicists discover previously unknown tectonic movements. Here, a cool blob of rock (blue) beneath Manchuria appears to extend under the hot blob (red) beneath Japan. This joins with other recent evidence suggesting there is a plate boundary in the northern Sea of Japan.

In the last ten years, Dziewonski says, seismic tomographers have convinced other geophysicists of the validity of their technique. Now tomographers have assembled the best set of maps yet, and the time has come to compare the maps in detail with the theoretical models of the mantle's behavior being developed by theorists like David Yuen. Seismic tomography can at last be used to test and modify these simulations. To do so, researchers must pore over seismic images, looking for structures that may or may not be explainable by the models.

"At the December meeting of the American Geophysical Union in San Francisco, results of modeling and tomography were shown side by side," Dziewonski says. "That's essentially what some of us always thought we could do, but now it's actually happening."

Yuen's model and others have shown that the Earth's mantle is not static. It is churning like a pot of boiling water on a stove. This motion comes about because of the difference in temperature between the mantle's upper and lower surfaces. Beneath the mantle is the Earth's hot core, above it the relatively cool crust. Heat flows from the core to the crust, but mantle rock is not a good enough conductor to carry all this heat without moving. Therefore, the modelers conclude, currents of rock must rise and fall, carrying heat from core to crust.

This process is analogous to what happens in boiling water. Some of the water (or rock) on the bottom of the pot (or mantle) gets hotter than its surroundings. It begins to rise, displacing some of the colder material above and causing it to move down. The red blobs in Morin's images correspond to rock moving up through the mantle, and the blue blobs are sinking rock.

The analogy isn't perfect. Many geophysicists believe that the convection of the mantle may be influenced less by heating from the bottom than by cooling from the top. The mantle may also receive some of its heat input from the decay of radioactive material within the mantle rock itself, not just from the hot core. Not enough is known about the properties of the mantle material to say which factor has the most significant impact on convection. One of the goals of the seismographers is to shed light on this question.

Another crucial difference between the mantle and a pot is the fact that the mantle is divided into two layers, the upper mantle and the lower mantle. Although early tomography studies revealed that the division occurs at 400 miles below the surface, geophysicists don't really know what kind of division it is. Are the two layers made of chemically different kinds of rock? Or are they chemically the same but arranged in different crystal lattices, like graphite and diamond? Dziewonski favors the latter view and believes that tomography may help resolve the issue.

Geophysicists also don't know precisely how the division of the mantle affects convection. Dziewonski describes theoretical modeling of convection in a divided mantle that Paul Tackley of the California Institute of Technology presented at the American Geophysical Union meeting. It showed how cold blobs sinking from the top may accumulate in pools in the transition zone between the upper and lower mantle. Every half billion years, the model showed, the pools of cold rock break through all at once and sink to the bottom.

Dziewonski believes seismic tomography shows how the model could be refined. His tomographic images do indeed show pools of cold material forming in the transition zone — but they also show more cold material in the lower mantle than Tackley's model can account for.

"This is where things like tomography get very important," Dziewonski says. He believes that blobs from the upper mantle must break through every 10 million years or so to account for the number of cold blobs seen in his tomographic maps.

Tackley doesn't regard Dziewonski's point as a serious criticism. Because so little is known about the properties of the mantle, Tackley used very rough estimates for many of the most important variables in the model.

"I think it's very encouraging that the seismic tomography shows that there is pooling at the transition," he says, "and I don't think that the difference in timing is that important at this stage." As knowledge of the physical properties of the mantle improves, so will the models.

One of the great strengths of the theory of mantle convection is that it provides a motor for the Earth's tectonic plates. The movement of continents across the face of the Earth is actually part of the mantle's convective current. In fact, Yuen, Dziewonski and many other geophysicists believe subduction of one tectonic plate beneath another drives the entire convective cycle. When two plates collide, one plate is forced under the other. When the subducted plate — the one that is forced down — reaches the mantle, it becomes a cold blob and sinks. Then as it moves down through the mantle, it displaces hotter material, which eventually rises to the surface in regions where plates are moving away from each other.

One such place is Iceland, where the North American and Eurasian plates are moving apart, leaving an ever-larger North Atlantic between them. Hot mantle that rises to fill the gap gives rise to the island's volcanoes. Seismic tomographic maps, including Morin's, clearly show a large hot blob directly underneath Iceland.

Morin's map of the eastern Pacific Ocean reveals another part of the plate tectonic picture — one that has only recently come to light. The map shows a cool blob beneath Manchuria that appears to extend under a hot blob directly beneath Japan. This seems to confirm the suspicion of some geophysicists that Japan and Manchuria are not both on the Eurasian Plate. Manchuria and part of the seafloor are on a separate plate that is subducting beneath Japan at a boundary in the northern Sea of Japan.

Yuen found another surprise while exploring Morin's visualization. Jutting down 125 miles below southeastern Missouri, Yuen says, is a toothlike spike of cold rock.

The spike is almost exactly under the New Madrid fault zone, the site of three massive earthquakes in 1811 and 1812, whose magnitudes have been estimated as ranging from 8.1 to 8.3. (See "New Madrid: The Rift, the River, and the Earthquake," *Earth*, January 1992, p. 34.) Researchers have long wondered why an area in the middle of the solid North American Plate, far from the plate's boundaries where quakes ordinarily occur, would have such large temblors. When Yuen saw the spike, he immediately wondered whether it could be massive enough to pull down on the Earth's crust in the area, creating strains that could lead to quakes.

Several months of additional study have convinced him that his hunch was right. Yuen found that earlier researchers had identified a gravity anomaly — a region where the strength of the Earth's gravitational field is slightly weaker than normal — in the region. "In the past, it was thought that the gravity situation was due to glacial rebound," Yuen says. (Glacial rebound is the upward movement of the crust caused by the melting of millions of tons of ice at the end of the last ice age.) But now it's clear that the region is being pulled down from beneath by the spike of cold mantle rock, and this is causing the gravity anomaly.

While he was working on verifying his hunch about the New Madrid spike, Yuen stumbled across a similar structure under the east coast of Australia, another area far from tectonic plate boundaries. That site is also known for occasional earthquakes and scientists have struggled to understand why. "They thought it was sediment building up and pushing down on the region," Yuen says. "Now we believe [the area] is being pulled down from below."

Yuen has made understanding the age and origin of these spikes one of the goals of his research. He has developed a whole line of research from an observation he could never have made without seismic tomography — or without Paul Morin's interactive computer graphics. And that's after only one year: "Almost every time we look at this we find something new," Yuen says.

As for Morin, he has spent several intense years adapting software, developing computer procedures and learning enough about geophysics to work with the data. All the while his linguistics work has languished, just a few credits short of a degree.

"And now they've closed linguistics school," he says. "I guess I'll just have to go somewhere else." ⊕

Jim Dawson is a science reporter with the Minneapolis Star Tribune.

Photo by C. Heliker, U. S. Geological Survey

Hawaii's Volcanos: Windows Into The Earth

by John Dvorak

The volcanos of Hawaii are probably the most thoroughly studied in the world. They are helping Earth scientists get a better understanding of how volcanos on Earth, and on other planets, operate.

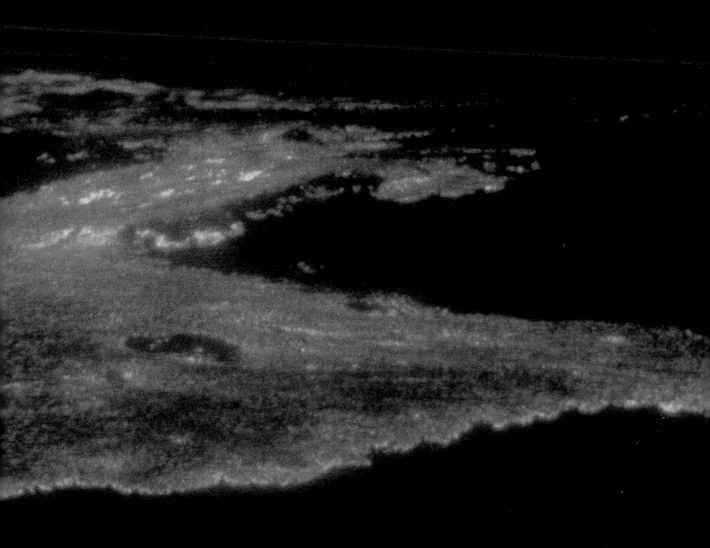

When freshly erupted (above), Hawaiian lava is rich in dissolved gases and flows easily in the form called pahoehoe (pah-HOY-hoy). Photo by J. D. Griggs, USGS. Fountain eruptions, driven by gases in the liquid rock, can reach heights of hundreds of feet. Photo by C. Heliker, USGS.

Volcanos are our most direct link to the inside of the Earth. They are a source for samples of material that normally lie deep beneath our feet, and they give us a feeling for the degree of restlessness inside the Earth.

More than 80 percent of Earth's surface has been formed by volcanic activity of one kind or another. Volcanos build mountains, form new islands, and spew dust and sulfur into the atmosphere. Of the several hundred active volcanos, about a hundred show signs of activity in any given year. Of these, a few produce major eruptions that damage property and may require nearby populations to evacuate.

Predicting the behavior of most volcanos is difficult because decades or centuries may pass between eruptions, and the buildup to an eruption may take place in only days or months. To study them, scientists have established permanent observatories at several of the world's more active volcanos. One of these observatories is on the island of Hawaii.

In part because they erupt frequently, Hawaiian volcanos are probably the most thoroughly studied in the world. Compared to explosive volcanos such as Mount St. Helens and others in the western United States, eruptions of Hawaiian volcanos are mild. This allows volcanologists to examine eruptions close up while keeping a few safe steps in front of advancing lava flows. Many studies of Hawaiian eruptions have given Earth scientists a good view of how this kind of volcano works. This, in turn, helps us understand what is happening inside our planet.

Scientists have learned that Earth is slowly cooling as heat escapes from the deep interior to the surface. In the rocky layer called the mantle, which underlies the outer crust that we live on, heat moves by convection. Rocks in the mantle are hot enough and under enough pressure to behave as a very viscous fluid. The result is that the mantle flows in a circular pattern with hot rock rising in some places and cooler rock descending in others. The overall pattern is like a conveyor belt, with the outer layer, the crust, riding passively atop the convection cells.

Crustal movement is achingly slow by human standards — only a few inches per year, or about as fast as fingernails grow. Mantle convection has broken the crust into many plates, each moving in a distinct direction. North and South America are each a single plate, as are Africa, Australia, and India. The ocean floor also is divided into plates. Most of the Pacific Ocean consists of a single plate moving northwest toward Japan.

The mantle's convection pattern is pierced in more than 100 places by rapidly rising plumes of rock called hot spots. The mantle is partially liquid near the top of many plumes and some of this liquid rock, called magma, erupts and forms volcanos. The chemistry of this magma differs from that erupted at a plate boundary, being richer in alkali metals such as sodium and potassium. This suggests that the source of hot spots lies a few thousand miles down, while crustal boundary magma is shallower in origin.

Hot spots remain stationary while crustal plates move over them. The result is a chain of volcanos drawn ac-

Shiny on the surface but soft and glowing with inner heat, toes of pahoehoe lava creep across black sand. Photo by J. D. Griggs, USGS.

cross the plate like a line of fire. This is how the Hawaiian Islands formed. As the Pacific plate moves northwest over the stationary Hawaiian hot spot, magma feeds a volcano until it is carried too far away. Then a new volcano grows over the hot spot until it also moves too far. Then a third volcano grows, and so on.

The Hawaiian hot spot has been supplying magma for more than 70 million years to a line of volcanos that march from Hawaii to the northwest edge of the Pacific plate (see pages 32-33). Older volcanos that now lie completely below sea level are called seamounts. The oldest in the Hawaiian chain is the Meiji seamount, located near the western end of the Aleutian Islands. Volcanos older than the Meiji seamount have disappeared, carried down into the mantle as the Pacific plate slides under the Aleutians. A sharp bend in the line near Midway Island records a change in the Pacific plate's direction of movement about 45 million years ago. The youngest and currently active Hawaiian volcanos are those at the southeast end of the chain.

That islands grow older northwest of the hot spot is shown both by the islands' relative state of erosion and by isotopic dating of rock samples. The crust has sagged under the weight of the volcanos. This, plus erosion by weather and sea waves, has lowered the height of all but the most recently active volcanos. Northwest of Kure Island, the summits of all Hawaiian volcanos are now below sea level. From Kure southeast to Niihau, the volcanos rise no more than a few hundred feet above the sea. You can also see progressive erosion among the seven major Hawaiian islands. Streams and waves have carved many deep valleys on Kauai and Oahu. Maui is eroded less, and only a portion of the north coast of Hawaii shows deep erosion.

Dating of rock samples can tell us the age of the last major volcanic activity for each island. Kure and Midway islands were last active about 28 million years ago. On Kauai, the major volcanic activity ended about 4 million years ago. Most of the surface lavas on Oahu were erupted between 2 and 4 million years ago, although small eruptions occurred a few thousand years ago. Among the features formed then is Diamond Head, the familiar landmark seen from Waikiki Beach. It is a broad volcanic cone formed when magma was erupted at a shallow depth beneath the sea. On Molokai, Lanai, Kahoolawe, and the western half of Maui, major volcanic activity last occurred from 1 to 2 million years ago. Maui consists of two volcanos: at the eastern volcano, Haleakala, the last major activity was 750,000 years ago, although one small eruption occurred only 200 years ago.

The five volcanos that comprise the island of Hawaii also show an age progression. Kohala, in the northwest, is the oldest; its last period of activity was 400,000 years ago. Mauna Kea, the site of many astronomical observatories, was active as recently as the last few tens of thousands of years. Hualalai in the west has erupted several times during the last few hundred years. Mauna Loa is building toward another eruption, and Kilauea is erupting now.

A volcano is considered active if it has erupted within the last few hundred years and dormant if it has erupted within the last several thousand years. By this standard, the Hawaiian islands have five active volcanos and at least one dormant one. The five active volcanos are Haleakala on Maui; Hualalai, Mauna Loa, and Kilauea on Hawaii; and Loihi, a seamount about 25 miles southeast of Hawaii. Mauna Kea, the one dormant volcano, has erupted at least three lava flows since glaciers disappeared from the summit at the end of the last ice age, about 9,000 years ago. It will likely erupt again, though the interval between active periods is on the order of several thousand years. Earthquakes occur beneath Mauna Kea: the most recent was a magnitude 4 event on September 1, 1990.

Although it appears as just a bump on the side of Mauna Loa, Kilauea is the volcano that erupts most frequently. Eruptions may be continuous for several decades, and during the past 40 years, they have occurred in many places on the volcano. About 10 percent of the volcano's surface has been covered since 1955 and 90 percent in the last 500 years.

Kilauea's current eruption cycle began in January 1983 along a segment of the volcano called the East Rift Zone. For the first three years, activity was intermittent, producing flows and lava fountains at monthly intervals. Beginning in mid-1986 the eruptive pattern changed. A lake of molten rock, continuously supplied by fresh lava from deep within the volcano, formed along the East Rift Zone. Eventually, a natural tube developed within the newest lava flows, which were fed by occasional spillovers from the lake. The tube runs about 10 miles down the flank of the volcano in an easterly direction toward the sea.

Occasionally, the lava flow in the tube is blocked, which causes lava break-outs somewhere on the surface. This led to the recent destruction of Kalapana Gardens, a small coastal community. During two weeks in April 1990, lava from Kilauea destroyed more than 60 houses there (see photos on page 34). Since 1983, more than 170 houses in several communities have been destroyed by Kilauea's lava flows.

Hawaii's Roots Go Deep

The island of Hawaii (the Big Island) stands over a plume of molten rock rising from deep in Earth's mantle. The plume punches through the crust to build the island by repeated eruptions from the magma reservoir beneath Kilauea. The Pacific Ocean crustal plate is moving northwest, however, so the plume has built a line of extinct volcanos and underwater seamounts reaching all the way to the Aleutian Islands.

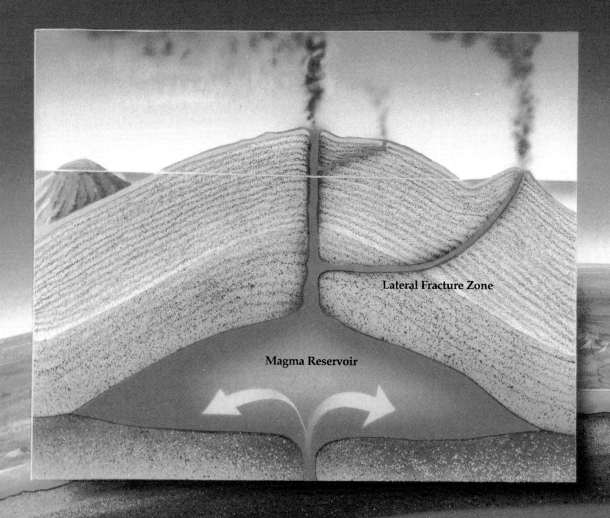

Painting by Paul DiMare

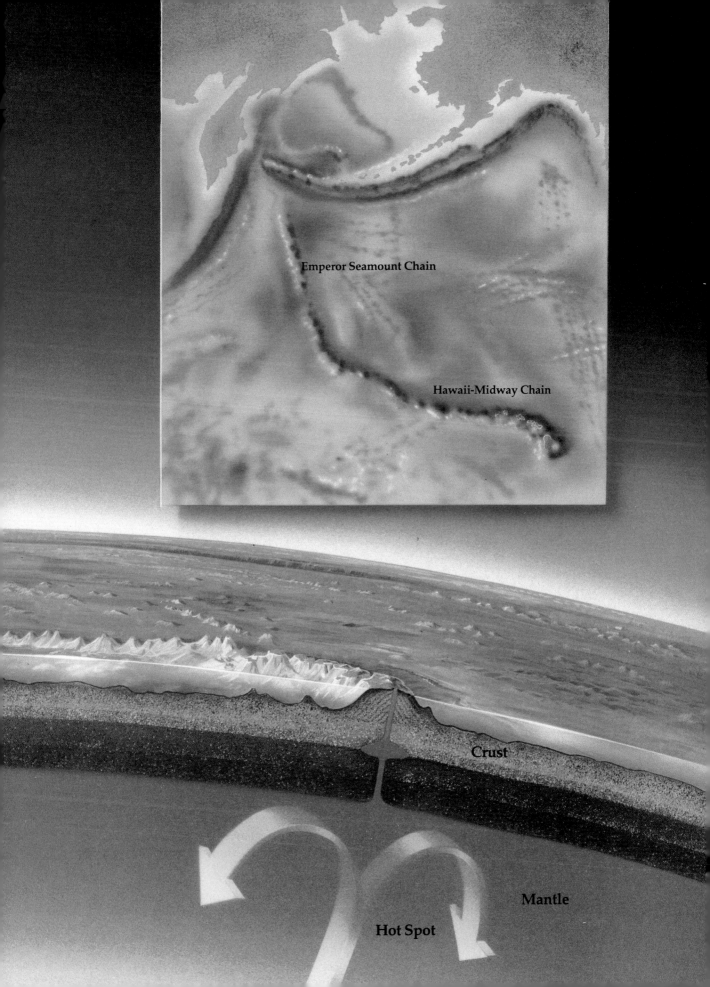

The community of Kalapana Gardens was destroyed last spring by lava: the top view shows the town on April 3, 1990. The middle view was taken May 2 and the bottom on June 3. Photos by J. D. Griggs, USGS.

The ropy folds of a pahoehoe flow are ideal for trapping wind-blown dirt and seeds, and life thus begins anew. Photo by J.D. Griggs, USGS.

The destruction of forests and houses is only part of the story. Hawaii's volcanos also create. The island of Hawaii is still growing as each successive flow adds another layer or extends the coastline: 160 acres have been added to the island since 1983, and the total area covered by post-1983 flows is about 30 square miles, equal to the area of Manhattan Island. Since 1986, lava has erupted at a constant rate of half a million cubic yards per day — a rate that would fill the Houston Astrodome, from playing field to ceiling, in about one week.

The much larger size of Mauna Loa compared to Kilauea means either that Mauna Loa is much older or that magma has been supplied in much greater amounts to Mauna Loa. When geologists compare the cumulative volume of lava flows since 1840, it appears that between 1840 and 1950 lava erupted three times faster at Mauna Loa than at Kilauea. But during the last 40 years the relative rates have reversed. However, the combined rate of lava erupted from both volcanos has been nearly constant since 1840.

Mauna Loa is by far the largest volcano in the world. If this seems surprising, remember that most of the volcano lies below sea level. Mauna Loa's summit reaches an altitude of nearly 14,000 feet, while the ocean nearby is about 15,000 feet deep. In addition, the weight of the five volcanos that comprise the island of Hawaii has buckled the Earth's crust, perhaps as much as 6 miles. If we add together the height above sea level, the depth of the ocean, and the amount of crustal sag, then the total stack of volcanic rock forming Mauna Loa may be as much as 12 miles thick.

While Kilauea is now Hawaii's most active volcano, it won't always be because the Pacific plate will eventually move it away from the hot spot. A younger volcano, named Loihi, is already forming beneath the ocean southeast of Kilauea. Though the summit of this youngest Hawaiian volcano rises more than 12,000 feet above

Fire meets water, and Hawaii grows a little larger. Already, the next Hawaiian island is being built underwater off the Big Island's southeast coast. A seamount named Loihi, it could emerge above the surf, according to scientists, in about 10,000 years. Photo by T. J. Takahashi, USGS.

the ocean floor, it is still 3,000 feet below sea level. If the hot spot feeds Loihi at the same rate as Kilauea and Mauna Loa, Loihi will reach sea level and become the next Hawaiian island in 10,000 to 100,000 years from now.

Geologists who have visited Loihi in deep-diving submarines have recovered fresh-appearing rock samples. Though no one has witnessed an eruption of Loihi, the fresh samples and the occasional earthquake swarms, most recently in March 1990, strongly suggest that Loihi is active. Detailed sonar maps of Loihi have been made in the last few years. When the surveys are redone, geologists may be able to identify new lava flows.

The best way to understand how a volcano works is to examine how it is constructed and how it behaves during an eruption. Kilauea and Mauna Loa have the same basic structure. The summit of each is dominated by a caldera, a large circular depression formed when magma withdrew from a reservoir that lies a few miles beneath the summit. As magma rises from the mantle, it accumulates in this reservoir, lifting the surface of the volcano slightly. When magma withdraws from the reservoir, the surface subsides. If a lot of magma is withdrawn, such as during a major eruption along the East Rift Zone, there is nothing left to support the summit rocks overlying the reservoir. They collapse and the caldera grows.

Magma collects in the reservoir over a period of several months to a few years before an eruption. Magma reservoirs are not unique to Hawaii's volcanos, but have been identified at many other active volcanos, such as Long Valley in California and Mount Etna in Sicily. As magma collects, the pressure in the reservoir increases, causing the number of earthquakes beneath the summit to grow. When the pressure finally opens new cracks or reopens older conduits, an underground pathway is formed for magma to flow from the reservoir into one of the rift zones.

The flow of magma into a rift zone is very rapid — and often spectacular when it results in an eruption. Magma can drain from the reservoir in a few hours. Geologists follow the flow of magma beneath the rift zone by tracking the location of earthquakes as they migrate. During one event in January 1983, earthquakes moved along the rift at a quarter-mile per hour. Eruptions begin with lava issuing from a long fissure, forming a continuous wall of lava fountains called the curtain of fire. These may be several miles long and jet lava hundreds of feet into the air. If the eruption continues for more than a few hours, the eruption will stay active at only a few points along the original fissure. By this time, the eruption may become episodic, with a single fountain shooting from a vent every few hours to a few months, sometimes to heights of over 2,000 feet. If the eruption continues, it finally forms a quiet lake of lava with little fountaining that may persist for decades. Kilauea's present lava lake has been active since mid-1986.

As magma flows into the rift zone, it forces the rift wider. At the beginning of Kilauea's 1983 eruption, the East Rift Zone widened eight feet in less than a day. As successive eruptions force more and more magma into a rift zone, the volcano's flanks compress much as a spring does. But rock elasticity has a limit, and eventually the stored energy is released by an earthquake. Several large quakes have occurred along the flanks of Hawaii's volcanos. On November 29, 1975, a magnitude 7.2 earthquake shook the south coast of Kilauea. This quake, slightly larger than the October 1989 earthquake in California, was the third largest to occur in the United States in the past century. The only two larger were the great 1906 San Francisco quake and the 1964 Alaskan earthquake near Anchorage.

Can Hawaiian volcanos tell us something about volcanos on other worlds? The question of how volcanos behave on Mars, Venus, Io, and elsewhere in the solar system will have to wait for future visits by spacecraft. However, two general conclusions are already evident. First, the major volcanos on Mars and Venus have the same overall shape as Hawaiian volcanos but lack well developed rift zones. The existence of summit calderas and, possibly, pit craters on the flanks of these volcanos probably indicate the existence of magma reservoirs.

Second, the largest Martian and Venusian volcanos are much larger in volume than Mauna Loa, Earth's largest. This is probably because the crusts of Mars and Venus are stationary and not broken up into moving plates, as on Earth. How long did it take to build a Martian volcano? Using the magma supply rate for Hawaii as a rough guide, the largest volcano on Mars, Olympus Mons, could have grown in 200 million years.

Volcanism is a major process in the evolution of rocky planets and moons. Until we can send an automated volcano observatory to other solar system bodies, our understanding of how all volcanos work will depend on studies of terrestrial volcanos — especially the most active, such as Hawaii's. □

John Dvorak is a volcanologist with the U. S. Geological Survey; he lives and works in Hawaii.

TEXT AND ILLUSTRATIONS BY BOB BAKKER

BAKKER'S FIELD GUIDE TO
JURASSIC PARK DINOSAURS

As Steven Spielberg prepared to make *Jurassic Park*, he narrowed an early list of more than 50 dinosaurs down to the 10 that would play parts in the film. He tried to be reasonably faithful to science in his portrayal of these animals, with one conspicuous exception: He doubled the size of *Velociraptor*, which played an important role in Michael Crichton's book. Since Spielberg's dinosaur was based on *Velociraptor* but closely resembles the then-undiscovered *Utahraptor*, we include both raptors in this guide. • What follows is a summary of the most up-to-date paleontology on the dinosaurs that provided the models for Spielberg's special-effects team, with field marks provided to help you recognize the dinosaurs as they appear in the film. Dinosaur sounds are inferred from the shapes of the sound chambers in the animals' skulls. Speed is estimated from the proportions of thigh, shin and ankle bones. Coloring is based on comparison with living animals.

COMPY
PROCOMPSOGNATHUS

A small predator, the dinosaur equivalent of a kit fox.

BASICS:

Time — Middle Late Triassic, 212 million years ago.
Length — 2 to 3 feet, including tail.
Weight — 10 to 20 pounds.
Speed — 35 to 40 mph in short bursts.
Voice — Chirps, burps, hisses and low grunts possible.
Habitat — Conifer forests and fern meadows.
Family — Podokesaurs, or "podokes" for short.

FIELD MARK: In *Jurassic Park*, compys have poisonous saliva, used to numb the face, hands and feet of prey. This is plausible: Some dino descendants (birds) have nerve toxins today, and a narcotizing bite has evolved in many other species, such as toads. • **SENSES AND INTELLIGENCE:** Big eyes that faced mostly sideways, providing little depth perception. Hearing less keen than a bird's in the high frequencies. Good sense of smell. Brain bigger relative to body weight than a lizard's but smaller than a bird's. • **GROUP DYNAMICS:** Travelled in packs, as indicated by fossil footprints. May have communicated by bobbing head, swishing tail and tilting torso up and down. • **FAMILY VALUES:** Probably laid eggs protected by both parents, as do most birds. May have nested in brooding communes. • **MANEUVERABILITY:** *Running* — bipedal, on hind legs alone. Flexible torso and tail permitted sinuous turns among tree trunks. *Climbing* — good, using both front and hind paws. Could grab branches with hand claws and curl hind paw toes around branches. *Digging* — good, using long, straight foot claws to break hard earth. *Swimming* — excellent, using hind kicks and tail strokes. • **ARMAMENT:** *Teeth* — small, sharp and back-curved, with saw edges on front and back. *Jaws* — small, slender and quick, good for fast nips. *Neck* — very slender, capable of delivering a fast snapping strike forward. *Hand claws* — small, with sharp tips curved. *Foot claws* — long, slender, almost straight, good for digging and kicking straight ahead. *Tail* — long and thin, useful as whip to stun enemies. • **ATTACK MODE:** May have formed foraging parties, scuttling through the forest underbrush stirring up small lizards, baby dinosaurs, furry mammals, big bugs and other assorted small fry. • **DEFENSE MODE:** Could scramble into trees or burrows to escape danger. • **WEAKNESSES:** Body, especially neck, very delicate and easily injured.

6 INCHES

Color rendering by EARTH: Phil Kirchmeier

RAPTOR
VELOCIRAPTOR

Wolf-sized predator, extremely vicious and deadly for its size.

BASICS:

Time — Late Late Cretaceous, 76 million years ago.

Length — 6 to 9 feet.

Weight — 60 to 140 pounds.

Speed — 35 to 40 mph.

Habitat — Mongolian species hunted in sand dunes and around little ponds in the desert. One North American species frequented dried-up lake beds and mud flats. Another American species preferred wet forests and swamps.

Family — Raptors.

1 FOOT

FIELD MARK: Raptors had huge, down-curved hind claws that could rip open the guts of victims ten times their size. • **SENSES AND INTELLIGENCE:** Big, hawklike eyes facing forward and sideways. Could focus both eyes directly to the front, providing excellent depth perception. Large chambers for sense of smell. Hearing good at middle and low frequencies. Brain huge for the body size, proportionately as big as brain in modern ground-running birds. • **GROUP DYNAMICS:** Probably hunted in packs. Although no fossil footprints of this family have been found, most other meat-eater trackways occur in groups. When attacking from all sides, raptors could bring down prey much larger than themselves. • **FAMILY VALUES:** Probably laid eggs and were highly protective of their nest and young, like most other meat-eating dinos. • **MANEUVERABILITY:** *Running* — strictly bipedal, with relatively huge hind legs and long but slender arms. Torso short and compact front to back, and rear half of tail very stiff, so body form was less sinuous and maneuverable in dense brush than a compy's. But stiff tail could be flipped up and down, side to side quickly, to shift balance when jumping or turning at high speed. *Climbing* — long, clawed fingers and hind toes, used to grip tree trunks and thick branches. *Digging* — weak, as only two hind toes on each paw had straight claws. *Swimming* — fair, but small hind-foot area and stiff tail meant lower speed than a compy's. • **OFFENSIVE ARMAMENT:** *Teeth* — small compared to jaw but sharp and curved sharply backward. Saw-toothed edges front and back. Most blade-shaped and narrow. Front tooth crowns blunt; front teeth arranged in a wide U-shaped pattern for biting off semi-circles of meat from prey. *Jaws* — very long and slender, not especially strong but quick. *Neck* — medium to short, but muscular. S-shaped curve of neck very sharp, so head and neck could pull jaws and teeth back and upward with great power. *Bite* — quick, with shallow tooth penetration and forceful backward jerk. *Hand claws* — gigantic. Down-curved claw tips like a hawk's, for holding onto prey (circular in cross section). High leverage of claw-flexing muscle, so hand claws could be driven deeply into large prey. *Foot claws* — second hind claw large and knife-shaped, with a strong downturn in the tip, the main killing weapon, slicing deep into the victim. High leverage for the claw-flexing muscle. *Tail* — rear half stiff, reinforced with many thin, bony rods, so tail tip did not make a good whip. • **DEFENSE MODE:** Could climb into low-hanging branches to escape dangerous enemies. • **WEAKNESSES:** Not good at moving through dense brush and not as fast at swimming as other predator families. Small hind feet meant difficulty in maneuvering in soft mud.

SPIELBERG'S GIANT RAPTOR
UTAHRAPTOR

Polar bear-sized predator with deadly claws fore and aft. Able to kill the largest dinosaurs in its habitat.

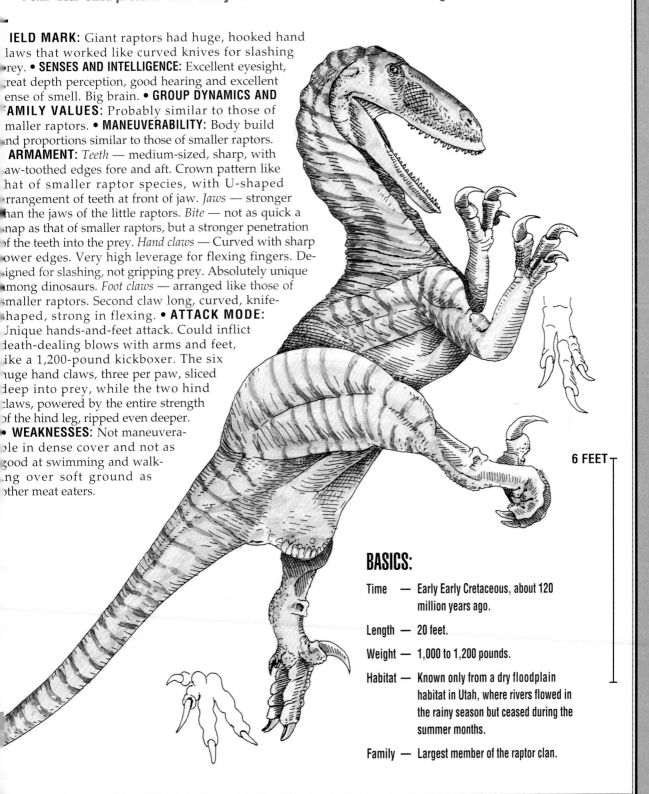

FIELD MARK: Giant raptors had huge, hooked hand claws that worked like curved knives for slashing prey. • **SENSES AND INTELLIGENCE:** Excellent eyesight, great depth perception, good hearing and excellent sense of smell. Big brain. • **GROUP DYNAMICS AND FAMILY VALUES:** Probably similar to those of smaller raptors. • **MANEUVERABILITY:** Body build and proportions similar to those of smaller raptors. **ARMAMENT:** *Teeth* — medium-sized, sharp, with saw-toothed edges fore and aft. Crown pattern like that of smaller raptor species, with U-shaped arrangement of teeth at front of jaw. *Jaws* — stronger than the jaws of the little raptors. *Bite* — not as quick a snap as that of smaller raptors, but a stronger penetration of the teeth into the prey. *Hand claws* — Curved with sharp lower edges. Very high leverage for flexing fingers. Designed for slashing, not gripping prey. Absolutely unique among dinosaurs. *Foot claws* — arranged like those of smaller raptors. Second claw long, curved, knife-shaped, strong in flexing. • **ATTACK MODE:** Unique hands-and-feet attack. Could inflict death-dealing blows with arms and feet, like a 1,200-pound kickboxer. The six huge hand claws, three per paw, sliced deep into prey, while the two hind claws, powered by the entire strength of the hind leg, ripped even deeper. • **WEAKNESSES:** Not maneuverable in dense cover and not as good at swimming and walking over soft ground as other meat eaters.

6 FEET

BASICS:

Time — Early Early Cretaceous, about 120 million years ago.

Length — 20 feet.

Weight — 1,000 to 1,200 pounds.

Habitat — Known only from a dry floodplain habitat in Utah, where rivers flowed in the rainy season but ceased during the summer months.

Family — Largest member of the raptor clan.

GIANT OSTRICH DINO
GALLIMIMUS

Ultra-fast running predator that hunts small reptiles, eggs, furry mammals.

BASICS:

Time — Late Late Cretaceous, 69 million years ago.

Length — 15 to 20 feet.

Weight — 300 to 600 pounds.

Speed — 50 mph.

Habitat — Mongolian species found in sand dunes. Close kin found in swampy forests and meadows in Canada.

Family — Ornithomimids, the ostrich dinos.

3 FEET

FIELD MARKS: Ostrich dinos had super-fast hind-leg action, speedier than that of any other predatory dinosaur. Also had tiny, toothless heads set on long, thin necks. • **SENSES AND INTELLIGENCE:** Eyesight excellent, with forward-directed eyes giving great depth perception. Hearing slightly better than usual for a dinosaur. Sense of smell good but not as well developed as in other meat eaters with larger snouts. Brain huge for the body size, as big as brain of a ground-running bird today. • **GROUP DYNAMICS:** Foraged in large adult gaggles, as indicated by tracks. Group hunting helped flush out small prey and provided early detection of tyrannosaurs and other predators. • **FAMILY VALUES:** May have nested in large groups. • **MANEUVERABILITY:** *Running* — strictly bipedal. Rear half of tail stiff, like a *T. rex*'s, so turning in dense forest difficult. *Climbing* — could not climb. *Digging* — excellent, using long fore claws to break hard soil and dig out prey. *Swimming* — good, using kicks of hind legs. • **ARMAMENT:** *Teeth* — none. *Jaws* — weak, with small biting muscles. Jaw edges made into sharp-edged beak. *Neck* — thin and flexible. Head bent down on neck very sharply. *Bite* — like an ostrich's, a lightning strike of head and neck down and forward. *Hand claws* — long, almost straight. Three hand claws of almost the same length, forming a rakelike excavating device, a unique design. *Foot claws* — short and blunt, moderately curved. OK for kicking forward but not deadly in effect. *Tail* — too stiff to be a good weapon. • **ATTACK MODE:** Gaggles of ostrich dinos probably fanned out over a meadow, searching for small prey hiding in burrows, holes and crevices. Dug prey out quickly with fore claws and then snatched it with a head-neck strike. *Escape tactic*: Could run away at high speed. When cornered, could kick with hind legs and lash out with fore claws. • **WEAKNESSES:** Neck and head easily damaged. Couldn't accelerate quickly if caught in dense brush.

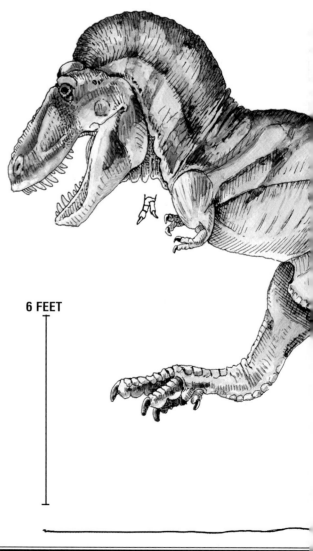

6 FEET

T. REX
TYRANNOSAURUS REX

Gargantuan top predator, heaviest and most dangerous meat eater of its time, as massive as an elephant and capable of killing any plant eater with one bite.

FIELD MARKS: Body mass was concentrated into two functions: running and biting. Forehead very wide, jaw muscles and neck muscles immensely strong, producing a bite-and-shake action twice as strong as that of any other predator. • **SENSES AND INTELLIGENCE:** Eyes small compared to head size, but both eyes could focus forward over the narrow snout for excellent depth perception. Outer ears widely spaced by broad head, with ear openings facing forward, giving superb depth perception in hearing too. Large muzzle allowed strong sense of smell. Brain of moderate size for a dinosaur, far smaller than a raptor's in comparison to body bulk. • **GROUP DYNAMICS:** Probably hunted in small groups, as indicated by footprints. • **FAMILY VALUES:** Males probably fought annual courtship battles featuring head butting at full speed, as indicated by blunt bony growths developed to make a helmet around eyes. Broken and healed ribs in *T. rex* fossils suggest violent ramming. Cheekbones below eye often scarred by bite marks, showing that ritual combat incorporated biting at the face. • **MANEUVERABILITY:** *Running* — strictly bipedal. Rear half of tail stiffer than a compy's, but not quite as stiff as a raptor's. Tail used for quick flips right or left, up or down, for turning while running in open habitat. *Climbing* — much too big to climb. Small, weak front legs prohibited climbing even when young. *Digging* — good, using large hind paws with straight claws to excavate prey from burrows. *Swimming* — strong swimmers, using kicks from hind legs, but slower than other carnivores with more sinuous tails. • **ARMAMENT:** *Teeth* — huge, strong, armor-piercing crowns, thicker than usual for meat eaters, more like railroad spikes than knives. Saw-toothed edges front and back. Crowns able to smash through ribs, shoulder blades and skulls. *Jaws* — most massive jaw muscles of any meat eater. High leverage for biting, yet bite remained fast because of long muscle fibers. *Neck* — incredibly wide and deep, giving tremendous strength for shaking prey. Neck cocked into tight S-shaped curve, making neck muscles stronger for pulling prey backward. *Bite* — fast and very strong. *Hands* — small, with only two fingers. Claws blunt and not strongly curved. Weak leverage for finger-flexing muscles. Weak muscles for pulling arm in and back. *Foot claws* — long, straight and facing forward, useful for kicking but not nearly as deadly as a raptor hind claw. *Tail* — thick and muscular at base, thin and stiff aft. Not long and supple enough to make a good whip. • **ATTACK MODE:** Struck swiftly in pairs, running much faster than any potential prey. Depth perception important in aiming lunges of head and neck. Super jaws could crush through the neck or thorax of a *Triceratops* in one bite. • **WEAKNESSES:** Could not maneuver in dense forest or bush as well as four-footed prey like the horned dinosaurs. Weak arms weren't much use in holding prey while biting.

BASICS:

Time — Latest Cretaceous, about 66 million years ago.

Length — 44 to 50 feet.

Weight — 3 to 5 tons.

Speed — 35 to 40 mph.

Habitat — Adapted to wet forests, brushland and swamp borders in western Canada and the United States.

Family — Tyrannosaurs.

DILOPH
DILOPHOSAURUS

A large predator, the equivalent of a fast, limber polar bear.

BASICS:

- Time — Early to middle Early Jurassic, 190 million years
- Length — 20 to 25 feet.
- Weight — 1,000 to 1,500 pounds.
- Speed — 33 mph.
- Voice — Limited to the hiss-chip-roar range heard today alligators.
- Habitat — Warm, frost-free conifer forests and floodplains
- Family — Podokesaurs.

FIELD MARKS: Dilophs had two tall head crests, side by side, for threatening enemies and attracting mates. In the book and movie, dilophs also spit nerve poison, like spitting cobras. • **SENSES AND INTELLIGENCE:** Standard dino eyes with color vision, facing outward, so eyes can't focus dead ahead and depth perception limited. Superb sense of smell. Hearing good but not as sensitive to high frequencies as that of modern birds. Brain size moderate. • **GROUP DYNAMICS:** Footprints of big meat-eating dinos usually occur in small groups, so dilophs probably hunted in twos, threes or fours. Crests must have been used in dramatic head-bobbing, weaving and thrusting. • **FAMILY VALUES:** May have nested as mated pairs, with both male and female guarding eggs. • **MANEUVERABILITY:** *Running* — strictly bipedal. Long, sinuous torso and tail design allowed quick turns and easy travel through dense brush and forests. *Climbing* — too heavy to climb trees. *Digging* — could excavate burrows of prey with hind feet. *Swimming* — swift, powered by big hind feet and very long flexible tail. • **ARMAMENT:** *Teeth* — long, slender and back-curved, with saw-toothed edges front and back. Front teeth long curved spikes for grabbing prey (round in cross section). *Jaws* — extremely long, slender and fast-snapping. Not especially strong but excellent for seizing prey. *Neck* — shorter and thicker than a compy's, not as quick in flexing and extending but stronger for shaking prey. *Hand claws* — medium-sized, gently hooked, but not as strong and curved as raptor claws. Short, strong front leg for pulling prey back and inward. *Foot claws* — useful as weapons when diloph rocked back on tail, as kangaroos do. *Tail* — like a compy's, very long, flexible. A good whip. • **ATTACK MODE:** Probably used scent to detect distant prey, as present-day long-snouted meat eaters do and then used both sight and scent to get in close. Used tail whip to stun small prey and confuse large victims. Snatched and swallowed small prey whole. Bit big prey with a quick snap of the teeth and a backward jerk of the head, leaving a long gash. • **WEAKNESSES:** Bite not especially strong — far weaker than *T. rex*'s.

100 EARTH

OTHY
DRINKER

Tiny, lively tree-climbing leaf eaters.

FIELD MARK: Othys were the most delicate plant-eating dinosaurs around, fast runners and good climbers, able to scamper across swampy ground and up into trees. • **SENSES AND INTELLIGENCE:** Huge eyes able to focus forward for good depth perception. Ears of usual dinosaur capability. Small snout with moderate organ of smell. Brain moderately large. • **GROUP DYNAMICS:** Probably traveled in gaggles of a dozen or more, as indicated by fossils found in large groups. • **FAMILY VALUES:** Large, communal nesting, with nests probably built by piling mud and vegetation into mounds on islands in the swamp. Groups probably fed together and warned each other of danger. • **FOOD PROCESSING:** Cut tough leaves and branches with coarse saw edges on teeth but couldn't chew toughest leaves. • **MANEUVERABILITY:** Hind legs three times as long as forelegs, thirty times stronger. Torso stiff because of bony tendons around backbone; end of tail stiffened the same way, so body not sinuously flexible. *Running* — on hind legs only, but forepaws could be used for walking slowly. *Climbing* — good, using all four feet. Grasped branches with inner toes of big, spreading hind feet. *Digging* — excellent, using straight claws on hind feet. *Swimming* — fast, using strokes of hind feet. • **ARMAMENT:** *Back teeth* — sharp and triangular, with coarse saw-tooth edges. *Front teeth* — long with fine saw-tooth edges. *Jaws* — short but strong. Jaw muscles slow acting. *Neck* — thin, weak and cocked into an S-shaped curve. *Bite* — left short but nasty, jagged-edged wound. *Hand claws* — very short, blunt and down-curved, not good weapons. *Foot claws* — very long and very, very sharp, arranged in a three-pronged stiletto. Extremely dangerous in kicking to the front. Inner claw sharply down-curved, nasty weapon for gouging. *Tail* — too stiff to be a good whip. • **DEFENSE MODE:** Counter-attacked in groups, with quick hind kicks, plus sudden scatterings of entire gaggle into the swamp or up trees. • **WEAKNESSES:** Very tiny and very weak, vulnerable neck; no weapon to keep predator at a distance.

BASICS:

Time — Latest Jurassic, 140 million years ago.
Length — 3 to 4 feet.
Weight — 10 to 20 pounds.
Speed — 35 mph.
Habitat — Found only in or near swamps.
Family — Othnielians.

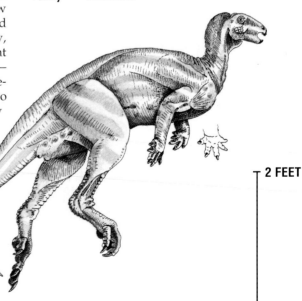

2 FEET

BRONTO
BRONTOSAURUS
(ALSO CALLED APATOSAURUS)

Stupendously large vegetarian, feeding on tree tops.

FIELD MARK: Anatomy of hips and backbone indicate brontos could rear up on hind legs and push heads 35 feet high. Their whiplike tails had tremendous speed and reach and could cut an enemy in half. • **SENSES AND INTELLIGENCE:** Eyes large, facing sideways. Hearing about average for a dinosaur. Sense of smell moderate. Brain absolutely ridiculously tiny compared to body weight. Amount of memory storage capacity small, compared to that of most meat eaters. • **GROUP DYNAMICS:** Travelled in tight herds of thirty or more, young and adults together, as indicated by footprints. • **FAMILY VALUES:** Young always found with adults. Smallest young too bi[g] in an egg, so probably born alive and kicking. Young spec[ies] rare compared to adults, so infant mortality must have [been] low. • **FOOD PROCESSING:** Cut off branches and leaf cl[usters] with teeth. Ground food with hard rocks embedded in lin[ing of] huge gizzard. Extracted nutrients and water from grou[nd] plants in voluminous intestine. • **MANEUVERABILITY:** [Turn]ing — Normally four-footed. Could turn in dense forest [by] knocking down trees if necessary. *Climbing* — impossible. [Dig]ging — good, using three blunt hind claws to find water [and edi]ble roots. *Swimming* — strong, using hind-leg kicks an[d tail] strokes. • **ARMAMENT:** *Teeth* — long, pencil-shaped, p[acked] close together to make a cookie-cutter apparatus. *Jaws* — [small] for the size of the body; jaw muscles puny compared to a [meat] eater's. *Neck* — long and flexible, with immense hollow [verte]brae surrounded by quick-acting muscles. Entire neck [could] swing as a blunt, flail-like weapon. *Bite* — very weak for [its] size, not a usable weapon. *Hand claws* — only one, o[n the] thumb, very blunt, curved inward. Not a good weapon. Body bulk so large, the bronto could crush enemies under forepaws after rearing up. *Foot claws* — three on inner toes, all blunt, curved outward. Not good weapons. *Tail* — an exceptional weapon. Rear thirty feet or so a supple whiplash, front part massively muscled for side-to-side flexing. Whip speed exceeded 300 miles an hour. • **DEFENSE MODE:** Herds kept vulnerable young protected behind a screen of adults. Standing on hind legs and tail, they could scan for predators over the treetops. Mighty swings of whiptail could keep predators away. • **WEAKNESSES:** Not smart. Could get bogged down in swamps and soft mud because compact feet didn't spread weight far enough.

BASICS:

Time — Latest Jurassic, about 145 million years ago.

Length — 90 to 100 feet.

Weight — 20 to 30 tons.

Speed — 20 mph tops.

Voice — Honking call like a large woodwind instrument.

Habitat — Most common in floodplain forests with severe dry season.

Family — Diplodocids.

10 FEET

BRACH
BRACHIOSAURUS
Super-tall muncher of tree tops.

BASICS:

Time — Late Jurassic, about 145 million years ago.

Length — 100 feet.

Weight — Up to 40 tons.

Speed — 20 mph tops.

Voice — Loud, mooselike bugle.

Habitat — Most common on seasonally dry floodplains, near big rivers and temporary ponds.

Family — Brachiosaurs. Not closely related to Brontosaurus.

FIELD MARK: Brachs had high shoulders and extremely long necks, enabling them to browse tree-top leaves 45 feet above ground. • **SENSES AND INTELLIGENCE:** Eyes, ears and sense of smell on par with those of brontos. Brain tiny. • **GROUP DYNAMICS:** Travelled in large, tightly knit herds. • **FAMILY VALUES:** Protected young within herd. • **FOOD PROCESSING:** Crushed and cut branches and leaves with teeth. Ground food in gizzard. Had intestines even larger than those of brontos, so they could thoroughly digest tough, dry leaves. • **MANEUVERABILITY:** *Running* — Four-footed most of the time. Front legs much longer than hind, but hind legs twice as strong as front so brachs could rear up if they wanted to, though not as easily as brontos could. *Climbing* — impossible. *Digging* — good, using hind paws. *Swimming* — slow, since tail small and feet so compact. • **ARMAMENT:** *Teeth* — with thicker crowns than those of bronto but not as closely packed in the mouth. *Jaws* — strong but slow-acting. *Neck* — long but not nearly as strong as that of bronto. *Bite* — not nearly strong enough to be a good weapon. *Claws* — too blunt to be dangerous, but brachs could crush enemies under their front feet. *Tail* — short and weak, not a good weapon. • **DEFENSE MODE:** Could detect meat eaters from far away because of great height. Couldn't be as aggressive as brontos because tail and neck so much weaker. But sheer size made adults dangerous. Could stomp on unwary predators and crunch them. • **WEAKNESSES:** Not smart. Lacked a whiptail or any other weapon to keep predators at a distance. Could get bogged down in soft mud.

9 FEET

T-TOPS
TRICERATOPS

Elephant-sized bush eater, very dangerously armed.

BASICS:

Time — Late Late Cretaceous, 69 million years ago.

Length — 18 to 25 feet.

Weight — 3 to 6 tons.

Speed — 35 mph.

Voice — Courtship notes super-loud but duller in tone than those of a Brontosaurus.

Habitat — Commonest in wet forest, rare in dry habitats.

Family — Close kin of Torosaurus, Pentaceratops, Chasmosaurus; more distant kin of Styracosaurus.

FIELD MARK: T-tops had massive legs both front and rear, a compact, quick-turning torso and exceedingly deadly horns. • **SENSES AND INTELLIGENCE:** Eyes facing outward, with no forward or backward visual field. Ears the usual for a dinosaur, facing outward. Large snout space for sense of smell. Brain small for body size. • **GROUP DYNAMICS:** May have travelled in large herds, since horned dinosaurs are sometimes found in bone beds with 100 or more specimens. • **FAMILY VALUES:** Nesting grounds not yet found. • **FOOD PROCESSING:** Chopped up thick, tough leaves and branches with cutting cusps of powerful molars. Lacked gizzard stones. • **MANEUVERABILITY:** *Running* — normally on all fours but could rear up. Torso short, stiffened by bony rods, but strong front legs allowed quick turns and fast back-up speed. *Climbing* — good, using long toes, but T-tops too heavy to go up vertical slopes or trees. *Digging* — excellent, using wide hind paws. *Swimming* — fast, dog-paddling with wide front and hind paws. • **OFFENSIVE-DEFENSIVE ARMAMENT:** *Teeth* — no front biting teeth, molars too far aft to do damage. But beak sharp-tipped and potentially as dangerous as that of a 3-ton snapping turtle. *Jaws* — extremely strong but couldn't open wide. *Neck* — very strong, muscular, ball-in-socket joint at head-neck junction permitted quick head thrusts in every direction. *Bite* — could sever a tyrannosaur's shin. *Horns* — long horn over eyebrow, pointing upward, outward and forward, plus shorter horn over nose. Short, outward-facing horn below eye on each cheek. Short, sharp hornlets all along the frill edge plus three small horns on the midline of the frill, behind the eyes. Short horn, facing backward on the base of each big eyebrow horn. *Horn thrust* — deadly, could kill any dinosaur with one thrust. Frill hornlets could cause a long, jagged wound when head was swung. • **DEFENSIVE ARMAMENT:** *Hand and foot claws* — too blunt to be weapons. *Tail* — very thin and weak; not a good weapon. • **DEFENSE MODE:** Counter-attacked with short bursts of speed, fast turns and quick flips of the skull. • **WEAKNESSES:** No forward or backward vision.

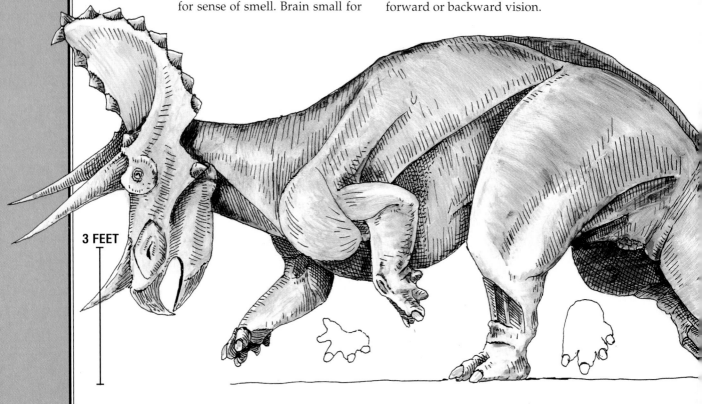

3 FEET

TROMBONE DUCKBILL or PARA
PARASAUROLOPHUS
Large, rhino-sized leaf and branch eater.

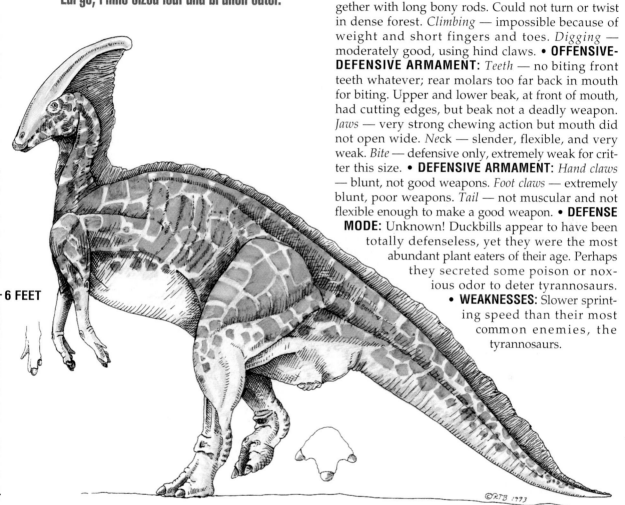

6 FEET

Hypacrosaurus, show breeding grounds were crowded with many family groups in a small area. • **FOOD PROCESSING:** Cut bunches of leaves or ripped off branches with wide beak. Had best molars of all dinosaurs for chewing hard, gritty plants. No gizzard stones. • **MANEUVERABILITY:** *Running* — used hind legs or all fours, but hind legs ten times as strong as forelegs, so most of thrust was from the rear. Torso and most of tail stiff; backbone bound together with long bony rods. Could not turn or twist in dense forest. *Climbing* — impossible because of weight and short fingers and toes. *Digging* — moderately good, using hind claws. • **OFFENSIVE-DEFENSIVE ARMAMENT:** *Teeth* — no biting front teeth whatever; rear molars too far back in mouth for biting. Upper and lower beak, at front of mouth, had cutting edges, but beak not a deadly weapon. *Jaws* — very strong chewing action but mouth did not open wide. *Neck* — slender, flexible, and very weak. *Bite* — defensive only, extremely weak for critter this size. • **DEFENSIVE ARMAMENT:** *Hand claws* — blunt, not good weapons. *Foot claws* — extremely blunt, poor weapons. *Tail* — not muscular and not flexible enough to make a good weapon. • **DEFENSE MODE:** Unknown! Duckbills appear to have been totally defenseless, yet they were the most abundant plant eaters of their age. Perhaps they secreted some poison or noxious odor to deter tyrannosaurs. • **WEAKNESSES:** Slower sprinting speed than their most common enemies, the tyrannosaurs.

FIELD MARK: Trombone duckbills had six-foot, backward-projecting, hollow horns on heads, providing gigantic echo chambers for making courtship songs. • **SENSES AND INTELLIGENCE**: Eyes huge and set on outwardly projecting eye sockets, covering almost 360 degrees; good for detecting predators in open terrain. Ears of usual dinosaur sensitivity, facing outward. Brain small for body size, with large sensory region for smell. • **GROUP DYNAMICS:** Sometimes found in huge bone beds with hundreds of individuals of all ages; probably traveled in big herds. • **FAMILY VALUES:** Nests of a close relative,

BASICS:

Time	Late Cretaceous, 74 million years ago
Length	30 feet.
Weight	2 to 4 tons.
Speed	35 mph or more.
Voice	Adults with very loud and complex courtship calls, like gigantic oboes. Babies with high-pitched, squeaky voices.
Habitat	Forests and seasonally humid meadows, avoids long dry seasons.
Family	Hadrosaurs, the hollow-crested duckbill dinos.

Extinctions — or, Which Way Did They Go?

by Steven M. Stanley

...mpact of a large meteorite or a comet 66 million ...ago may have killed off the dinosaurs and much ...life by altering the world's climate severely for the .... But some groups, including these pterosaurs, were *...bly already on the wane. Painting by Ron Miller.*

Life's evolutionary road has encountered many tortuous turns and dead ends. Large fractions of all creatures have died from time to time. While an asteroid impact is probably to blame for the death of the dinosaurs, most other extinctions are linked to fundamental geological processes — and carry a message as up to date as today.

Evolution, life, and the global ecosystem just aren't what they used to be. During the past decade, Earth scientists, life scientists, and even space scientists have been falling over each other in uncovering surprising new revelations about the mass extinctions that from time to time have blown a dark fog across periods of Earth's history. This exciting enterprise has invigorated all of geology. But except for the single event that swept away the dinosaurs, the new findings have remained unknown to the general public.

To begin with, the great extinction in which the dinosaurs died is just one of about a half-dozen major extinctions that have taken place in our planet's 4.6-billion-year history — and it is by no means the most dramatic. (The biggest we know of may have killed more than 9 out of every 10 living animal species, some 250 million years ago.)

Sheer curiosity might prompt a scientist to ask what caused these catastrophes. But every human being can't help also wondering, "How vulnerable are we?"

These questions, unfortunately, are easier to pose than to answer. What causes a mass extinction? It appears there may be no single answer. In many cases, a change in climate conditions was the culprit. But few scientists would be content with that — we need to know what drove these changes. And so we dig ever deeper into the past, finding answers that seldom deliver as much certainty as we'd like!

The Demise of the Dinosaurs

It would be exaggerating a bit, but not too much, to say that the modern history of mass extinctions began in 1980, when a team of workers led by physicist Luis Alvarez and his geologist son Walter, made a crucial discovery that revived interest in the long-standing problem of what killed off the dinosaurs. They found an abnormally high concentration of the rare element iridium that was laid down in a thin layer of sediments in many parts of the world. This layer corresponded exactly to the time when dinosaurs died out at the end of the Cretaceous Period 65 million years ago.

The Alvarez team recognized that iridium is much more abundant in meteorites than in the Earth's crust. They suggested that the extinction of the dinosaurs and

many of their animal contemporaries resulted from the impact on Earth of a large meteorite or a comet — a bolide, as they called it. They calculated that the volume of iridium indicated a bolide with a diameter of about 10 kilometers (6 miles). Depending on whether it struck water or land, this would have left a crater about 100 to 125 kilometers in diameter.

Strong additional evidence found subsequently also pointed to an extraterrestrial cause for what is now called the terminal Cretaceous event. Chief among this was an abundance of minutely-shocked mineral grains in the iridium-rich zone. Found in many parts of the world, these grains are for the most part sand-sized particles that display multiple sets of shock deformations called lamellae. These are parallel planes of recrystallization visible with a microscope.

Some experts believe that explosive volcanic eruptions could produce grains of this type, but the only proven origin remains meteorite impact, which can generate shocks of enormous magnitude, with pressures measured in millions of atmospheres. The shocked mineral grains are largest and most abundant in sedimentary rocks from the American west, where there is also the best fossil record of the last known dinosaurs. It seems reasonable to expect that the size and abundance of shocked grains would decrease away from an impact site; this led to a suggestion that the bolide struck in the western United States.

There are no 100-kilometer craters visible there at present. (Arizona's well-known Meteor Crater is about 1 kilometer across and only 50,000 years old; see page

The Ordovician ocean of 450 million years ago contained swimmers (such as nautiloids), crawlers (trilobites), and relatively immobile creatures (brachiopods). All groups suffered greatly in the extinction at the end of the period. Smithsonian Institution photo.

50.) But buried underground at Manson, Iowa, is a circular geological structure that was discovered by well-drilling and which appears to record an impact. Radioactive minerals indicate an origin quite close to the time when the Cretaceous ended, about 65 million years ago. Is the Manson crater the site of the "killer" bolide? Maybe, but the Manson crater is only about 35 kilometers across.

Recently, Alan Hildebrand and William Boynton of the University of Arizona published what they regard as evidence that the impact occurred in the southwestern Caribbean Sea. Using seismic reflection profiles of the ocean sediments off Colombia, shocked minerals in core samples, and thick deposits of debris in Cuba and other areas, they claim to have identified a structure measuring about 300 kilometers in diameter as the impact crater. Is it actually? Perhaps. But other potential origins for this structure and the debris have not been ruled out, and it's much too early to declare the case closed.

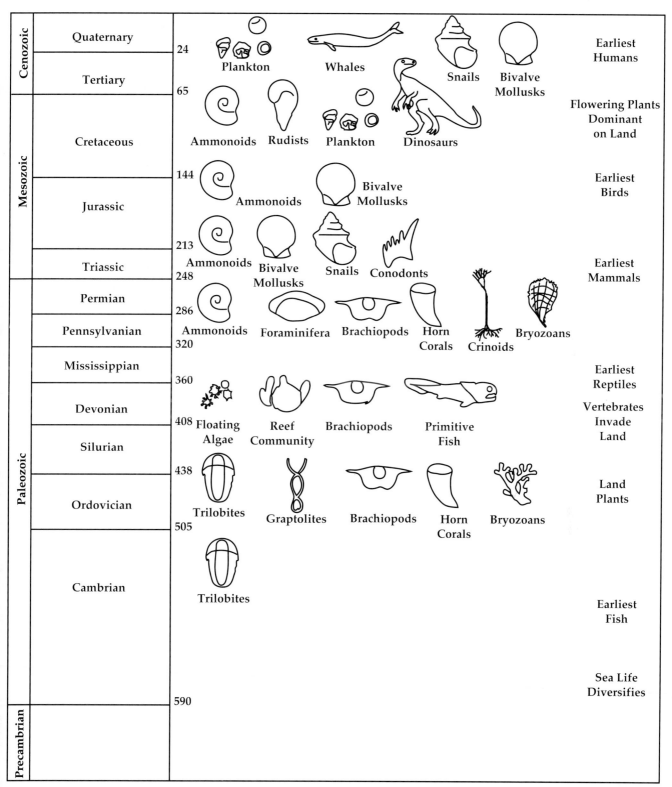

Even setting aside the question of crater site, the Cretaceous extinction picture has some puzzling details. For example, paleontologists are struck by the lack of evidence for sudden extinctions among terrestrial species in the Southern Hemisphere. In North America, fossil pollen and leaves of land plants reveal that from New Mexico to southern Canada, terrestrial vegetation underwent a sudden change at the same time that the iridium and shocked mineral grains were deposited. Many species of hardwood trees and other flowering plants died out, and primitive ferns briefly colonized the depopulated landscape. These ferns left

JANUARY 1991 109

vast quantities of their spores in the sedimentary record, producing a so-called "fern spike," before flowering plants reclaimed the land. But in the Southern Hemisphere, where shocked mineral grains are smaller and less densely concentrated, terrestrial floras of Cretaceous age appear to have given way gradually to those typical of the age of mammals that followed and in which we still live. This points to an end-of-Cretaceous event that is much less catastrophic than the impact model proposes.

Also puzzling is the fact that certain groups of marine organisms were declining well before the impact. For example, two or three million years before the end of the Cretaceous, the bivalve mollusks known as rudists were dying off in great numbers. The rudists — giant clams that built tropical reefs during the latter half of the Cretaceous — were in effect the "dinosaurs" of the mollusks. Although the last rudists may have died out with the dinosaurs, the group as a whole was waning long before. A million years before the rudists disappeared, their reefs were biotically impoverished and poorly developed. An unrelated group of giant bivalves, the inoceramids, died out altogether slightly before the Cretaceous came to a close. Do these facts negate an impact? No, but they may downgrade its role from being sufficient in itself to being simply the final blow.

The notion of a terminal Cretaceous impact naturally raises the related question of whether other major extinctions had a similar cause. These are even harder to evaluate than the Cretaceous event. After more than

In the chalk sea cliff at Stevns Klint, 25 miles from Copenhagen, is the Fish Clay, an iridium-rich layer (arrows) that many scientists think is the "smoking gun" of the impact that killed the dinosaurs. Photo by W. Crawford Elliott.

a decade of intensive work, the results remain stubbornly mixed. The story, frankly, is a tangled one with conflicting masses of circumstantial evidence.

It appears that extraterrestrial causes are feasible for some biotic crises, but not for others. Standing in the way of hard conclusions, however, are several areas of active dispute. These include agreeing on how one identifies a crisis as such, assessing how bad a given crisis was, determining if there is a likely impact candidate, and calculating how much damage an impact would inflict. Scientists even disagree on whether some catastrophes were actually severe enough to be called mass extinctions at all.

One big difficulty lies in determining how severe a crisis was. Scientists can't simply count the bodies because the fossil record is too incomplete. Estimates that use families (the next level of biological organization above species) are more accurate because families often include many species whose ranges in geologic time are relatively well known. Even so, there's an important caution to keep in mind: samples of families include only those whose members possessed hard body parts that were readily fossilized. Creatures made mostly of soft tissues are generally invisible in the rock strata. It's possible that many biological groups have come and gone unnoticed simply because the victims lacked hard body parts.

The most catastrophic extinction in the fossil record is the one at the end of the Permian Period, about 250 million years ago. This crisis, which ended the Paleozoic Era, may have swept away as many as 95 percent of all animal species on Earth. That, by any standard, was a gigantic catastrophe. Among its victims were most groups of mammal-like reptiles, the ancestors of mammals, and many kinds of marine animals. This catastrophe eliminated more than half of all marine families with good fossil records, and many families survived by the persistence of just a few species. (By comparison, only about 15 percent of all marine families died out at the end of the Cretaceous.)

What is a major extinction, anyway? Do regional or local pulses of extinction merge with global ones? How do these differ from the piecemeal, continual dying out of species all over the world which is termed background extinction? In other words, does a mass extinction simply intensify the ordinary processes that drive the background extinction rate?

Some mass extinctions exhibit curious ecological patterns. For example, the terminal Cretaceous event caused heavy losses among many kinds of phytoplankton (floating algae), but left virtually unscathed species that could go dormant when the environment turned hostile. By producing and occupying a protective cyst,

In the Cambrian Period over 500 million years ago, trilobites were a dominant group that scavenged the sea floor worldwide. Smithsonian Institution photo.

Trilobites suffered repeated mass extinctions during the Cambrian Era. In the graph below, each line represents a species. Note that many species lasted less than a million years. Diagram after S. M. Stanley.

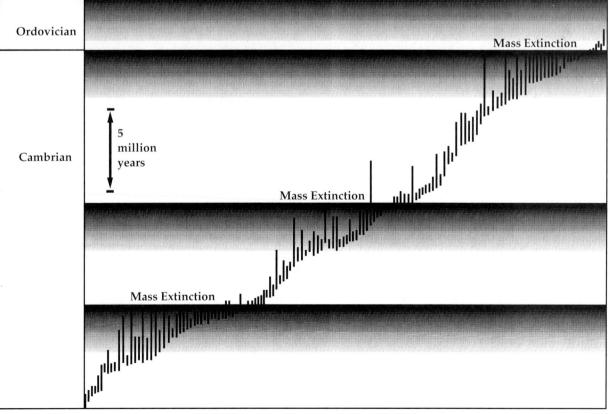

these species rode out the storm and emerged once conditions improved again.

In fitting impact scenarios to extinctions, a key issue is the frequency of impacts large enough to wreak havoc on a global scale. Of the 25 impact craters with diameters of 5 kilometers or more that occurred in the last 250 million years, only 4 exceed 50 kilometers' diameter. But the only crater known to be both younger than 250 million years and also roughly the size estimated for the terminal Cretaceous object is the 100-kilometer Popigai crater in the Soviet Union. Yet with an age of 30 million years, Popigai is far too young. How critical is this as a counterexample? Actually, not very — the record of known impacts is, like fossils, too incomplete. On average three bolides out of four will hit the oceans, and as the Manson crater shows, even on continents large craters can remain undiscovered or poorly dated.

Popigai, however, does challenge us to ask what the relationship is between bolide size (and speed) and the severity of biotic damage. Here is a big crater with no clearly associated extinction. Can otherwise identical impacts have vastly differing biological repercussions? Is it possible that a bolide caused heavy extinction at the end of the Cretaceous — but the Popigai bolide did not — only because Cretaceous life was already under stress? The decline of certain forms of marine life was noted already; it's also true that the youngest dinosaur faunas in the American west, although apparently rich in species, were dominated numerically by the horned herbivore Triceratops. (Dominance by one or a few species is the sign of an unhealthy ecosystem, or at least one that is precariously balanced and highly vulnerable to external influence.)

The record of marine plankton, best displayed in sediments deposited on continental shelves, suggests that there was indeed a sudden event (or several) at the very end of the Cretaceous Period. The record of heavy extinction just before the boundary raises the possibility that a comet shower or multiple impacts caused the crisis. Teams of geologists are currently re-examining key stratigraphic sequences where previous investigators have reported finding multiple iridium anomalies.

Other Crises, Other Causes

In hopes of clarifying the overall picture of extinctions, scientists are studying biotic crises with approaches that range from looking at what the chemistry of the sediments tells about conditions to carefully locating where species dwelt in geography and time. The impact scenario has spurred geologists to look especially closely at other mass extinctions in two different ways. First, they have scrutinized the microhistory of the rock record laid down during biological crises, seeking evidence of potentially lethal global changes. Second, they have examined the individual crises in search of patterns of extinction that point toward a particular cause.

Causes can be assessed at several levels. For example, evidence may suggest that climate change caused a particular mass extinction, but this leaves open the question as to what caused the critical change. And if the answer turned out to be that ocean currents had shifted profoundly, you then ask what caused the shift. (At times, working in this field can feel like trying to take apart one of those nesting Russian *matryoshka* dolls!)

At present, however, there is scant evidence that impacts caused any mass extinction other than the Cretaceous one. Ordinary geologic processes that can concentrate heavy metals appear to account for the few small iridium anomalies that have been found at other levels in the stratigraphic record. These anomalies occur only locally or fail to align stratigraphically with global biotic crises. Similarly, abundant amounts of shocked quartz grains are known from no crisis but the Cretaceous. It also appears that some mass extinctions have spanned millions of years and that some have occurred in pulses or intermittent waves.

For example, a look at the extinction event that took place during the late Eocene less than 40 million years ago is revealing because its pattern is well displayed in deposits on land and in ocean sediments sampled by deep-sea drilling. Terrestrial floras show that in Europe and North America the climate became cooler and drier during the latter part of Eocene time. Since the end of the Eocene, about 34 million years ago, climates at middle and high latitudes have been relatively cool and highly seasonal. (The severity of the change is shown by the fact that in early Eocene time, southeast England was cloaked in a forest like the jungles of modern Malaysia.) Terrestrial mammals also suffered two pulses of extinction, the second coinciding with the climate change at the close of the Eocene. Life in the seas experienced pulses of extinction as well, with tropical species suffering especially heavy losses; both floating organisms (plankton) and bottom-dwelling ones were affected.

What caused it? Geologists discovered that the late Eocene extinction coincided with a dramatic worldwide change in the ecology of the deep oceans. Life on the sea floor was transformed abruptly. Oxygen in the skeletons of deep-sea organisms changed toward higher concentrations of the heavier isotope, O^{18}, relative to O^{16}, showing that the water had become colder. This marked the formation of the psychrosphere, the name given to the zone of cold, deep bottom water that persists to the present day.

Then as now, this water mass was formed primarily by a giant natural refrigeration system near Antarctica. Some of the waters from the counterclockwise gyres of the Pacific, Atlantic, and Indian Oceans become trapped in the great clockwise current that encircles Antarctica. The water loses heat, thereby increasing in density, and sinks to the ocean floor with a temperature only slightly above freezing. This frigid water then spreads northward along the bottom throughout all of the major ocean basins. But here and there, this cold water can rise to the surface in upwellings that occur where winds push warm surface waters away from coastlines.

It thus appears that the climate changes that caused extinctions at the end of Eocene time are linked to events near the South Pole. There is independent evidence that Australia, which had been attached to An-

Two successful ammonite genera that disappeared with the dinosaurs at the end of the Cretaceous Period were Scaphites (above) and Turrilites (below). Smithsonian Institution photos.

tarctica for hundreds of millions of years, broke away at this time and began to drift northward toward its present location. This initiated the present water-chilling system, although an underwater ridge between Antarctica and South America prevented the circumantarctic current from developing fully for a few million years. Exactly how events in this region could cool the climates at mid-latitudes remains uncertain. But it appears likely that the events of this interval relate to climate changes that were global in scale, occurred in pulses, and resulted at least in part from ordinary geologic processes.

Also attributable to purely terrestrial causes is one of the more severe mass extinctions: the crisis that ended the Ordovician Period about 440 million years ago. This event eliminated more than 20 percent of marine families that left fossil remains. The rock record indicates the extinctions represent a protracted crisis that spread over several million years and were linked to known geographic changes.

Plate tectonics may have been the forcing agent: during Ordovician time, the land masses that today constitute Antarctica, Australia, South America, Africa, and peninsular India formed a supercontinent called Gondwanaland. As the Ordovician progressed, Gondwanaland moved southward toward the South Pole. At the margin of the continent, in what is now the Sahara region of Africa, huge glaciers spread over the land, depositing boulders, gravel, and sand.

The glaciers caused a wide-reaching climatic chill, with profound effects on biotic communities. As families of species began to disappear, floating animals and

JANUARY 1991 113

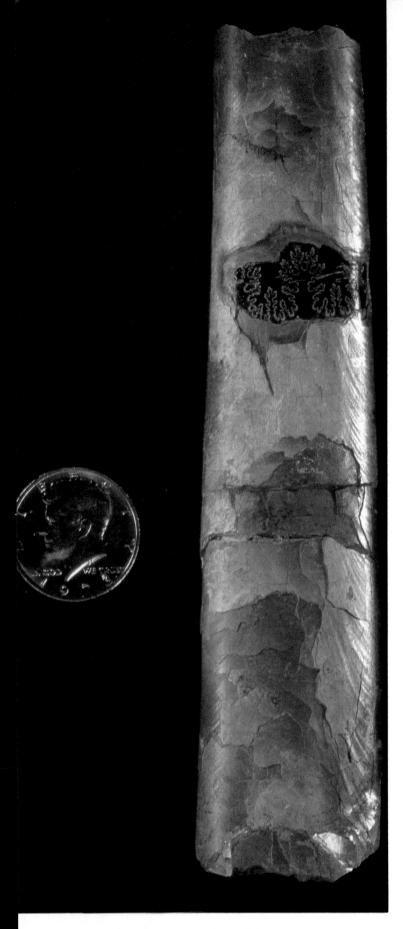

The straight ammonite Baculeites did not survive past the end of the Cretaceous. Part of this specimen's shell has been removed to show the convolutions in its internal partitions. Smithsonian Institution photo.

bottom-dwellers that lived in discrete latitude zones in the Ordovician seas moved toward the warmer equator. Sea level also dropped as great volumes of water were locked up in the glaciers. Some paleobiologists believe that the lowered sea level reduced the area of continental shelves, which could have contributed to the heavy extinction by reducing the room available for species to live in.

As with the Eocene event, the "terminal Ordovician crisis" can best be explained in terms of geological events, and in fact the pattern of extinction would be difficult to reconcile with an extraterrestrial cause like an impact.

How Often Do Bad Times Happen?

Are extinctions random in time, or is there a pattern? Two scientists at the University of Chicago, David Raup and John Sepkoski, have noted that extinction crises during the past 250 million years don't appear willy-nilly, but recur about every 26 million years.

Inaccuracies of dating, however, greatly complicate efforts to test for periodicity. Dates of events that occurred between 100 and 200 million years ago are subject to errors as great as 5 or 10 million years. Even the relatively recent end of the Eocene has been re-dated to 34 million years ago, some 3 or 4 million years younger than previously believed. This change expands the interval between the event and the terminal Cretaceous crisis to about 32 million years and shrinks the next interval to about 23. Previously these intervals had been considered approximately equal, spanning approximately 26 and 27 million years, respectively.

Another problem with periodic extinctions is that they explain too much. Adherence to periodic timing favors a common cause for all mass extinctions — and an extraterrestrial agent at that, because scientists know of no large-scale periodic terrestrial events. It is also difficult to reconcile periodicity with the clear evidence that links extinction events such as those of the Ordovician and Eocene to major geographic changes.

Finally, what we actually see isn't necessarily a true periodicity, but a pattern in which extinctions seldom follow one another in close succession. This, however, can be explained by supposing that after a major extinction there is generally a recovery time that makes it unlikely that another similar crisis will occur for at least 10 or 15 million years. New species do not appear instantaneously after a crisis, even if the environment returns quickly to its previous state.

We have a good example of delayed recovery in late Eocene time for a group known as the planktonic foraminifera. These creatures are single-celled floating organisms whose tiny skeletons rain down on the deep sea floor, leaving an excellent fossil record. Forams, as they are called, tolerate only small changes in temperature and the distribution of fossil species tells something about the temperature of the water. The late Eocene foram extinctions removed species that needed the warm, equable climates of low and middle latitudes. During the ensuing Oligocene Epoch, when climates were cooler and more variable, foram species remained few in number and were mainly ones that could toler-

late Devonian reef community decimated in the mass extinction that occurred toward the end of [the] period. We see nautiloids [straight and coiled], trilobites, cri[noid]s on wavy stalks, and colonies [of h]orn corals. Smithsonian Institu[tion] photo.

ate a wider range of temperatures. It took about 15 million years after the extinction for marine climates to become more equable. This allowed more narrowly adapted (and therefore vulnerable) species to evolve once again in large numbers. The next peak of extinction, albeit a minor one, occurred 7 or 8 million years later, when high species diversity had been reestablished.

What About Today?

As the record of the Ordovician and Eocene crises shows, there is abundant geological evidence that climate change has played a prominent role in mass extinctions. This can come about through changes in atmospheric or oceanographic currents, such as those of late Eocene time that redistributed heat within the oceans and atmosphere. Also possibly important are changes in the composition of Earth's atmosphere, which could vary the degree of greenhouse warming. Significant processes here are the liberation of oxygen by living plants and the consumption of carbon dioxide by chemical weathering of rocks and loose sediments at Earth's surface. Decaying plant debris yields methane, a greenhouse gas less notorious than carbon dioxide but as potent in its effects. Methane production has operated at a varying rate over geologic time, increasing, for example, when swamps expand, as they did during the Carboniferous coal age 280 to 360 million years ago. And, of course, a bolide impact could trigger sudden climatic change by blasting small particles into the upper atmosphere. Solid particles might screen out sunlight, cooling the Earth for a time. When these settle, droplets of liquid might remain in suspension, enhancing the greenhouse effect and warming the planet.

Future progress in understanding Earth's biotic crises will come from detailed case studies of particular events. Among the fruits of this ongoing research are insights into the fragility of our present global ecosystem, which is again under threat today, this time by the activities of its most intelligent species.

As we witness the stress to which our civilization is subjecting the rich, but fragile, biological communities on the savannahs of Africa and in the rainforests throughout the world, we should stop to remember how delicate the balance can be. The fossil record shows again and again that tropical communities repeatedly suffer most in major extinctions, largely because their species tend to be narrowly adapted and intolerant of environmental change.

It's the only planet we have. We ought to walk a little more lightly on it. □

Steven M. Stanley is professor of geology and chairman of the Department of Geological Sciences at Case Western Reserve University in Cleveland.

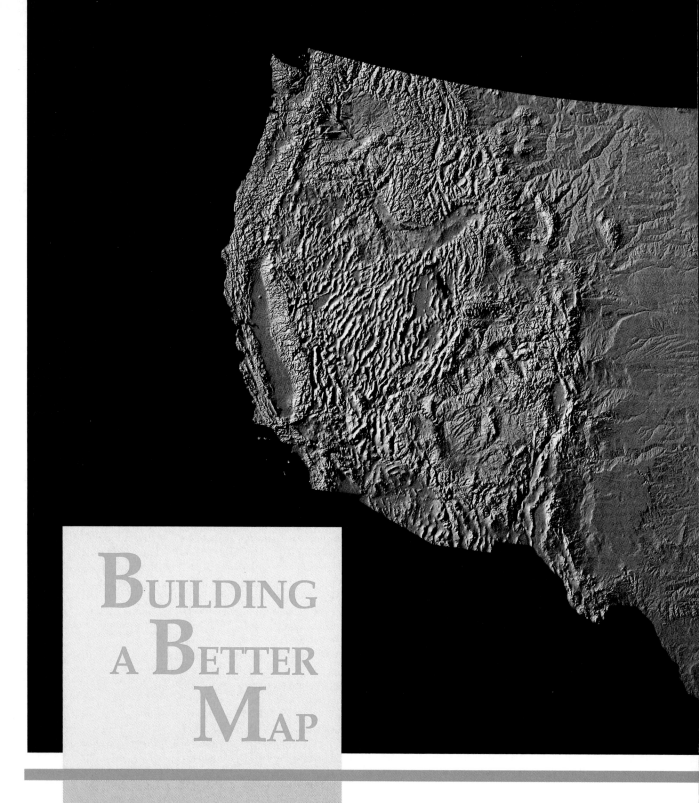

Building a Better Map

Maps are valued not so much because they express the obvious but because they depict the subtle.

by Richard J. Pike and Gail P. Thelin

When cartographers draw maps, they interpret the lay of the land symbolically through icons, color, and shading. Such maps may be beautiful and show large, obvious features well, but they are not realistic. We have, however, relied on such maps throughout most of our history.

That is changing.

Today, the use of computer graphics reveals features and patterns as never seen before. With unprecedented clarity and detail, we view phe-

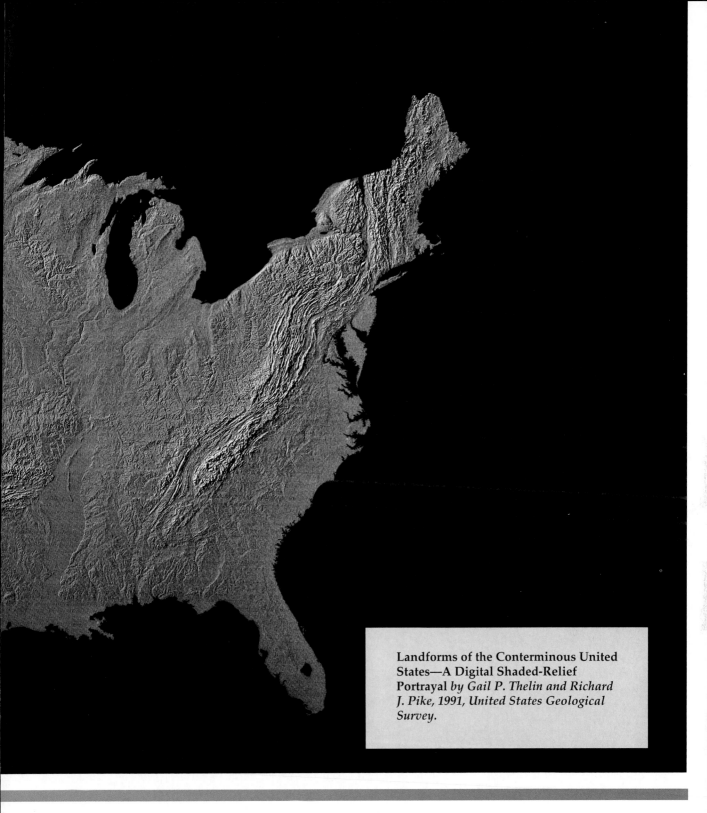

Landforms of the Conterminous United States—A Digital Shaded-Relief Portrayal by Gail P. Thelin and Richard J. Pike, 1991, United States Geological Survey.

nomena that are too abstract, too complex, too remote, or — especially in the case of Earth's surface — too large to be visible to the unaided eye. Other methods of portraying topography, whether hand-drawings from field notes or photographs from space, do not accommodate *both* broad coverage and accurate detail. Machine visualization does.

The new map, *Landforms of the Conterminous United States*, is an example of what is possible with current technology. In the late 1980s, we began adapting digital image-processing techniques previously developed to manipulate spacecraft telemetry. We used existing land-based USGS topographical data. *Landforms* is a computer-shaded image calculated from a large matrix of terrain heights called a digital elevation model; twelve million such elevations provided the database. The breadth of its coverage is roughly 3 million square miles (almost 8 million square kilometers), while its 0.5 mile (0.8 km) resolution bestows exquisite detail. We are treated to a view at once stunning and precise.

As you examine *Landforms*, you quickly notice the dominating contrasts. Compare the tectonically static East coast of the United States with its passive continental margin to the active West coast where the North American and Pacific crustal plates collide and slide past each other. Dissimilar tectonic regimes show clearly: the Atlantic and Gulf coasts, which developed on a gently sloping continental shelf, differ markedly in detail from the Pacific coast, where the offshore profile is steep.

A somewhat subtler contrast is that between the areas on either side of the ice limit. (See map on pages 48-49.) The Ice Age profoundly affected the country's landscape. Scars of scraping and scouring, along with meltwater effects and characteristic till deposits, bear testimony to glacial history and enable geologists to trace the edges of the last advance of glacial ice. Topography north of the ice limit is muted compared to the jagged texture typical of unglaciated terrain to the south. The Driftless Area in southwestern Wisconsin, a place untouched by the most recent glaciation, stands out sharply from the surrounding drift-mantled and ice-altered terrain. The erratic drainage patterns you see north of the drift border are evidence of advances and retreats of the last ice sheet.

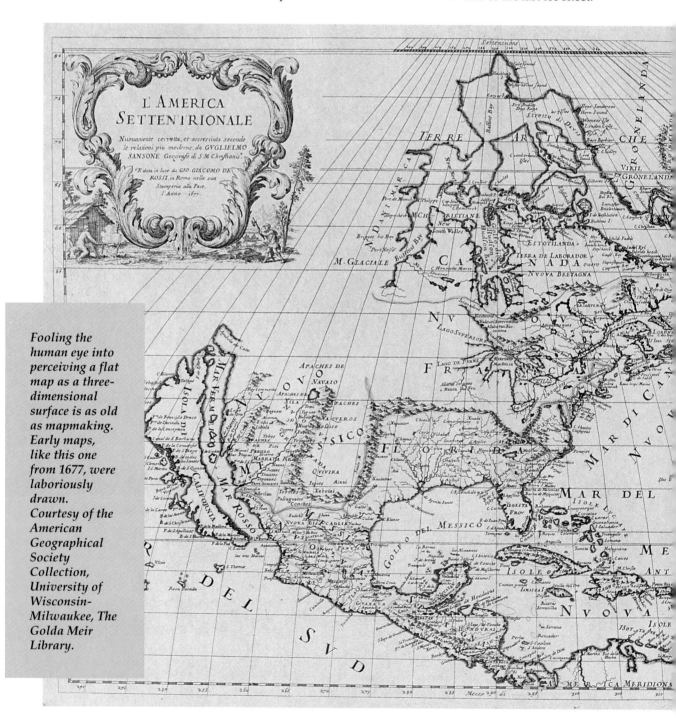

Fooling the human eye into perceiving a flat map as a three-dimensional surface is as old as mapmaking. Early maps, like this one from 1677, were laboriously drawn. **Courtesy of the American Geographical Society Collection, University of Wisconsin-Milwaukee, The Golda Meir Library.**

No less demanding was the classic work of Erwin Raisz. His maps have been used for decades by geologists and educators. Copyright 1954. Raisz Landform Maps, Jamaica Plain, Massachusetts.

Trace for yourself rock structures like the finely folded Ridge and Valley province of the Appalachian Mountains and the west-tilted Sierra Nevada fault block. Put your finger on the uplifted dome of the Black Hills that somehow escaped extensive warping and folding. Ponder the roughness and complexity of the Rocky Mountain system and the Pacific Border province. Scrape your knuckles on the grand washboard coarseness of the Basin and Range province, viewed in its entirety in detail previously unavailable at this scale.

Locate the filigreed Fall Line, named for falls and rapids of the Delaware, Potomac, James, Roanoke, Savannah, and other rivers flowing from the upland to the ocean. The Fall Line sharply divides dramatically different rocks of the Piedmont and Coastal Plain provinces.

The sculptured midcontinent may surprise you. Consisting largely of the Great Plains and Central Lowland provinces, this area contains landforms that are fully as remarkable as the mountain chains to the east and west.

Perhaps the most striking midcontinent formation is the Coteau des Prairies. The ice-scoured lowlands that flank the flatiron plateau in eastern South Dakota and neighboring states were occupied by the James and Des Moines lobes of the last ice sheet. The plateau and these lowlands drained meltwater from various ice-dammed lakes during deglaciation of the region.

An eye-catching feature of lower relief is the broad extent and extraordinary flatness of the Mississippi Alluvial Plain, the accumulation of soils and sediment carried and deposited by the meandering river. The Staked Plains of western Texas and eastern New Mexico also stand out. The fine-grained hummocky texture of the Nebraska Sand Hills marks the largest sand dune area in the Western Hemisphere. Crowleys Ridge, a late Pleistocene erosional remnant in the Mississippi River Embayment, is near the epicenter of the 1811-1812 New Madrid earthquakes.

By now, your observation skills are finely tuned and you are ready to find locally important landforms. Some require more scrutiny than others! In the East look for basalt ridges in the Connecticut Valley, glacial moraines on Long Island and western Cape Cod, and sandy Trail Ridge in northern Florida. In the West, you see the many low volcanic shields on southern Idaho's Snake River plain, two large calderas — Valles in New Mexico and Crater Lake in Oregon — and the approximate trace of the California San Andreas fault zone. That outstanding "glitch" in California's Sacramento Valley is actually Sutter Buttes, an enigmatic volcanic neck that appears out of place.

Look carefully, especially in the West, and you can see extended linear features which may indicate events in the tectonic evolution of the North American continent. Some features are seen clearly for the first time: the west-trending alignment in the Rocky Mountains of southern Colorado, the east-northeast trend that aligns with the Garlock fault zone of southern California, and an east-northeast alignment that includes parts of the Gila and Salt rivers in Arizona and the Canadian River in Texas. The latter two trends parallel the Murray fracture zone and other inactive and unseen transform fault systems of the eastern Pacific plate. These subtleties become visible only because of the fine resolution of the map.

Major elongate features more familiar to geolo-

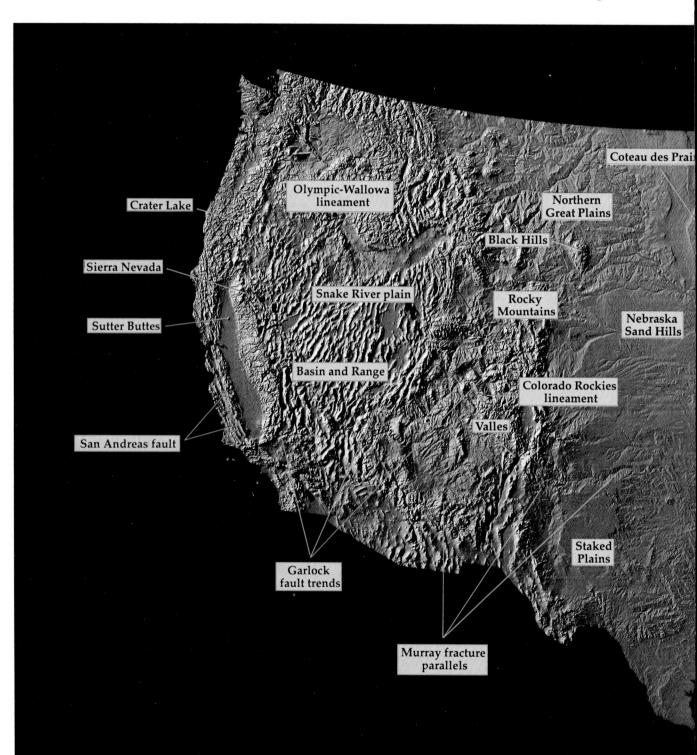

gists are Washington's Olympic-Wallowa lineament and the many, possibly related, northwest trends in the Rocky Mountains to the southeast. Some of these features, particularly the Olympic-Wallowa, may include currently active faults.

In some areas, minor lineaments may also indicate neotectonic activity. For example, the aligned stream segments of the northern Great Plains may reflect a pattern of etched bedrock fractures found throughout the middle one-third of the country, perhaps arising from the coupling of a mobile West with a more stable East. However, interpreting such trends requires restraint. The problem is this: The simulated low-angle sunlight used in producing the image artificially enhances small linear features, typically those 20° to 35° from the angle of illumination, while suppressing those in directions parallel to the source. For example, the selected 300° (west-northwest) lighting slightly exaggerates 325°-trending ridges and valleys and thereby enhances the distinct north-northwest grain of the Great Plains.

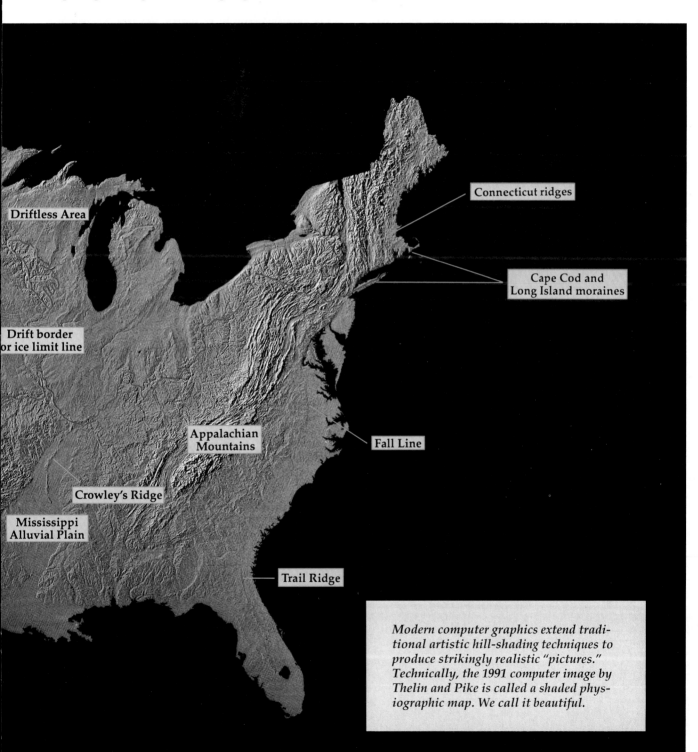

Modern computer graphics extend traditional artistic hill-shading techniques to produce strikingly realistic "pictures." Technically, the 1991 computer image by Thelin and Pike is called a shaded physiographic map. We call it beautiful.

Nonetheless, computerized relief maps offer significant advantages in the analysis of landforms. Above all, the mechanized shading portrays topography accurately and discloses its true complexity, allowing us to view discrete surface features in a broad regional context. Using similar computer techniques, stereo pairs in shaded relief can be created for an even more powerful three-dimensional effect. The Sun's position can be varied at will for different shadowing. Finally, images can be computed rapidly from digital files.

At first glance, digital shaded-relief maps look deceptively as if they are imaged from space. Compare the Lake Michigan area with its respective space portrait. Side-by-side, we easily see differences. On the new map, no clouds or vegetation mask elevation, no evidence of land use shows. It's as if we're seeing the face of the land scrubbed clean. Photo by NASA.

122   EARTH

Map accuracy jolts our sense of size and proportion. Mount St. Helens in southwestern Washington, impressively massive at close encounter, lies nearly hidden among the many larger peaks in the Cascade Range. The symmetry of the mountain reflects pre-1980 eruption topographical data. After seeing Mount St. Helens in this enlargement, try to locate it on the full spread at the beginning of this article.

This image of the United States was generated in 17 minutes, a tiny fraction of the time cartographers need to draw maps by hand.

Practical applications for computer-made relief maps abound. For topographic purposes, these include evaluating resources, mapping geologic hazards such as landslides, and interpreting regional and structural geology, global tectonics, and geomorphology. Shaded relief maps can be used as background maps for displaying other Earth science information such as climatic systems or earthquake epicenters and cultural data such as population density or transportation networks.

Like older maps, this new map is aesthetically appealing. The combination of broad scale and fine detail creates a beauty all its own. When you look at the map, you are almost transfixed by the land's amazingly intricate countenance. It is truly a map to savor.

Richard J. Pike is a geologist with the U.S. Geological Survey in Menlo Park, California. He "devoured" an atlas when he was eight years old and has loved maps ever since. Gail P. Thelin was with the USGS National Mapping Division at Ames Research Center, Moffett Field, California when this map was made. She is now a geographer with the USGS Water Resources Division in Sacramento, California. Thelin and Pike produced Landforms *because "it was fun" and "waiting to be done." Smaller areas had been similarly mapped. Faster computers and improved output devices made it irresistible to do the large expanse.* Landforms of the Conterminous United States—A Digital-Shaded Relief Portrayal *is now in print. Scale 1:3,500,000. Contact USGS Map Sales, Box 25286, Federal Center, Denver, CO 80225, for ordering information.*